Rabbinical Mathematics and Astronomy

RABBINICAL MATHEMATICS AND ASTRONOMY

BY

W. M. FELDMAN,

M.D., B.S., M.R.C.P. (Lond.), F.R.S. (Edin.)

Senior Physician St. Mary's Hospital. Plaistow.
Author of "Biomathematics" "Ante-Natal and Post-Natal Child Physiology, "Ante-Natal and Post-Natal Child Hygiene," "The Jewish Child," "Nursery Hygiene," "The Child," &c.

WITH PREFACE BY

R. A. SAMPSON, M.A., D.Sc., L.L.D., F.R.S.

Astronomer Royal for Scotland and Professor of Astronomy at the University of Edinburgh,

AND FOREWORD BY

ADOLF FRAENKEL, Ph. D.

Professor at Kiel University and Director of the Einstein Institute of Mathematics, Hebrew University of Jerusalem,

WITH FIFTY-SIX DIAGRAMS

HERMON PRESS

NEW YORK

Second, American edition

Library of Congress Catalogue Card Number: 65-15732

ISBN 0-87203-026-1

First edition: London, 1931

TO THE MEMORY OF MY FATHER

ISRAEL

WHO DIED AT THE AGE OF 76 YEARS,

ON WEDNESDAY, THE 9th OF MAY, 1928,

CORRESPONDING TO THE

19th OF IYAR, 5688

PREFACE

By R. A. SAMPSON, M.A., D.Sc., LL.D., F.R.S.

Astronomer Royal for Scotland and Professor of Astronomy at Edinburgh University; General Secretary and past Vice-President of the Royal Society of Edinburgh.

If a student of the history of mathematics and astronomy were asked what contributions to those sciences had been made by early Jewish scholars, he would, as a rule, have replied, that he could recollect none; and this would be surprising in view of the well-known and deeply significant examples of Greek and Indian thought, and the adventurous subtlety of more recent Jewish workers, of which some examples are mentioned in the last page of this work. One feels the necessity of exploring the ground further, and this Dr. Feldman's book permits one to do, for he combines apparently the learning, and certainly the industry, necessary for studying the Hebrew texts, with—what is equally essential for understanding and expounding them—a competence in mathematics. Any mathematician can now see why no advances were made by the Jews of past ages: they were interested in subtleties it is true, especially the subtleties of the Law, but not in those of Mathematics, nor of any natural science in which tradition counts for nothing. The regulation of the traditional Calendar forced upon them certain astronomical calculations which were competently performed, but the treatment, as any mathematician can see, is such as would lead to no advance.

I have read the book with pleasure, as the work of a fair-minded author, and it is with pleasure I commend it to anyone who cares for the out-of-the-way studies.

Royal Observatory, Edinburgh.
7th April 1931

R. A. SAMPSON.

FOREWORD

By ADOLF FRAENKEL, PH. D.

Professor at the University of Kiel, now Director of the Einstein Institute of Mathematics, Hebrew University, Jerusalem.

I am glad to meet the author's wish and to write a few lines as foreword to this book. The field dealt with therein holds much interest for Talmudists as well as for astronomers and mathematicians concerned with the history of these sciences. In view of the specific talent for mathematics and astronomy that modern Jewry is endowed with, and in consideration of the fact that there have been תלמידי חכמים skilled in mathematics, as well as mathematicians versed in Talmud, it is to be wondered at that the subject of this book was not developed comprehensively in recent times. The excellent works of Baneth, Zuckermann and others deal only with certain particular branches of the subject.

Thus it will be appreciated that through research devotedly carried on in a field professionally not his own, Dr. Feldman has given us a most extensive treatment of the subject. It is of no great importance that this book gives certain opportunities for criticism, or that one may have dissenting opinions concerning certain theses, in consideration of the fact that the appearance of this work presents a welcome starting-point for critique and further investigation.

I wish that, by its influence on large circles of readers, this book may again manifest the truth of the talmudical saying:

„כי היא חכמתכם ובינתכם לעיני העמים"—איזו היא חכמה ובינה לעיני העמים, הוי אומר זה חשבון תקופות וגמטריאות. (שבת ע"ה.)

Jerusalem ראשון לימי הקבלה (תרצ"א) / 19/5/1931

ADOLF FRAENKEL.

AUTHOR'S PREFACE

This work was primarily undertaken with the object of ascertaining, from an examination of those passages in the Talmud containing references to mathematical and astronomical matters, the extent of the knowledge possessed by the Talmudic Rabbis in the fields of Mathematics and Mathematical Astronomy, and the nature of any contributions they may have made towards the advancement of these sciences. The results of the investigation are briefly summarised in the last two pages of the book.

The research involved a close and critical study not only of the relevant passages themselves, but also of the annotations of the principal earlier Talmudic commentators, such as Rashi (11th century), the Tosefoth (12th and 13th centuries), and, above all, Maimonides, or Rambam (12th century), whose monumental work on the Consecration of the New Moon, *Kiddush Hachodesh*, has been subjected to a very minute and detailed analysis. The aim and character of that work are referred to on p. 111, and the various matters with which it deals are discussed in considerable detail in the various chapters of Part III. of this book. My study of that work—which has been greatly facilitated by the brilliant and classical Monograph of the late Professor Dr. E. Baneth, *(Maimuni's Neumondsberechnung)*— soon revealed the fact that whilst most of the numerical results given therein, are in agreement with those obtained by modern mathematical methods applied to Ptolemaic Astronomy, and are not widely different from those obtained from his data, by means of modern astronomical formulae, (which I give for comparison in various tables scattered throughout the book), others diverge so much from the calculated results that it is evident that some errors must have crept into the current printed editions of that work, in the course of transcription from the original manuscript. This led me to examine some of the available manuscripts of that work in the hope that I might discover the original one, written—or at least approved—by Maimonides himself. After examining the various MSS. in the British Museum, in Mr. D. S. Sassoon's private collection, in the Bodleian Library, Oxford, in the Frankfurt Stadtbibliothek, and in the Vatican Library, and after consulting Professor Dr. A. Freimann, Director of the Frankfurt Stadtbibliothek, I discovered that the original MS. is not in existence, but that the Oxford MS. of 1184 is the earliest to be found—it having been written during Rambam's life-time—although there is no evidence that it was revised by himself. It is this MS. which I have taken as the most authoritative.

All these MSS. differ among themselves, and whilst it is possible to explain some of the divergences, it is not easy to account for all of them. Thus, in the case of the Table on p. 146, the Lunar "Equation of the Centre" for an Anomaly of 40° is given in the Oxford MS. as 3°6′, which is in close agreement with calculation. The same value is given in the Vatican MS., but the British Museum MS. of the year 1335 gives the value of the "Equation" as 3°50′. It is obvious, therefore, that the copyist, in the course of his transcription, mistook the letter ו (=6) for נ (=50). On the other hand, the Sassoon MS. of 1355 gives the "Equation" as 3°44′, which can only be due to a slip of the pen, this being the correct equation for an anomaly of 50°. It is also possible that some of the errors were inadvertently made by Maimonides himself. For example, in the case of the Solar "Equation of the Centre" (see Table on p. 117), the correct value of the "Equation" for an anomaly of 170° is 0°21′, and this value is given in all the later MSS., but the value given in the Oxford MS. is 1°21′. If one bears in mind the numerous errors which constantly occur in the course of ordinary arithmetical work, it is astounding that such errors have been so exceedingly few, especially in view of Maimonides' extensive and multifarious duties as Rabbi, philosopher, author, Court physician, and very busy general medical practitioner. But after accounting for errors due to these causes, there still remains one Table of values in connection with lunar parallax (p. 151) in which the values given in different manuscripts, though in general agreement amongst themselves, differ so materially from those obtained by calculation, that it is impossible to explain the cause of those discrepancies.

Similar explanations apply to some of the Talmudic inaccuracies. Thus, in the description of the phenomena associated with twilight, quoted (from Sabbath 34b) on p. 92, it is fairly obvious that the words עליון (Zenith) and תחתון (Horizon) became transposed in the process of transcription, and the correct reading should be הכסיף העליון ולא הכסיף התחתון (" the zenith has greyed but not the horizon "), instead of הכסיף התחתון ולא הכסיף העליון, as given in the text, since during sunset it is the zenith which necessarily greys first. Indeed, Rashi, to Niddah 53a, actually gives the words in the correct order. Again, the figures 58,888 given in Berachoth 7a for the number of regaim in an hour (p. 188) instead of the correct number 82,080 are most probably an error inadvertently made by the copyist.

In addition to the academic character of the present work, I believe that the book will also be found to serve the purpose of an explanatory commentary to the various mathematical and astronomical passages to be found in the Talmud, as well as to Maimonides' *Kiddush Hachodesh*. **The Index to the references has been so arranged as to be of special help to the student in this direction.** It is, of course, not claimed that ' he who runs may read ' and follow easily the various points discussed in this book, but it is believed that the earnest student, who is accustomed to the concentration needed in the study of the Talmud and its various commentaries, and who possesses a rudimentary knowledge of Algebra and Trigonometry, will find it comparatively easy to follow the various steps of the mathematical manipulations and calculations given in this book.

A few words may perhaps be said regarding the state of astronomical knowledge during the Talmudic period. Although the ancient Chinese, Hindus and Babylonians were celebrated for their great astronomical skill and knowledge, the really scientific study of Astronomy began with the Greek Astronomer Hipparchus, who, in the 2nd century B.C. not only made numerous observations and discoveries, with the simplest possible appliances, but also invented the science of trigonometry for the purpose of measuring certain astronomical constants. For nearly 300 years after his death, there was no outstanding astronomical discovery made, but the teaching of Hipparchus gradually became disseminated throughout the world, and it is probable that much of it was acquired and adopted by the contemporary Talmudic Rabbis. In the middle of the 2nd century of the present era, Claudius Ptolemaeus, commonly known as Ptolemy, of Alexandria, wrote his great astronomical treatise, the *Almagest* which, amongst numerous other matters, postulates a *fixed*, *spherical* earth, round which the celestial sphere revolves carrying the celestial bodies with it (the Ptolemaic, or Geocentric system). There is no doubt that much of the information possessed by the Rabbis after the 2nd century was acquired from the Algamest, which was regarded as the astronomical Bible during the period of the 14 centuries that elapsed between Ptolemy and Copernicus. It was not, however, till the 4th century that R. Jonah accepted the theory of a spherical earth (p. 102), and it is also evident that the Talmudic Rabbis had very hazy ideas regarding the daily rotation of the heavens, although they ultimately adopted Ptolemy's geocentric theory (p. 71).

Of the shortcomings of this book I am well aware. Indeed, they are inevitable under the circumstances, inasmuch as I lack the ideal combination of extensive rabbinical scholarship with intimate knowledge of astronomy and expert skill in mathematics, necessary for the most adequate treatment of the subject. But, although I can adapt to myself the words of R. Nachman (Pesachim 105b) אנא לא חכימאה אנא ולא הוזאה אנא ולא יחידאה אנא for I am no Talmudic scholar, no practical astronomer, and no expert mathematician, I have made what use I could of what meagre equipment I possess in these branches of knowledge, and am gratified to know that, in the opinion of those able to judge, the result is not so imperfect as it might have been. Like R. Nachman, too, I have been a גמרנא וסדרנא, for I have made a systematic study of the works of previous writers, such as Zuckermann, Slonimsky, Mahler, Baneth and others, whose names I mention in the body of the book. I believe, however, that I have gone farther afield than any of these authors and that this volume is the first comprehensive book of its kind in any language.

It is probable that the captious critic may find in this book a certain amount of scope for cavilling, but I have confidence that the broad-minded scholar, whilst refusing to swallow a camel, will not strain at an occasional gnat.

I wish to record my indebtedness to the following gentlemen for various kindnesses shown to me during the progress of the book through the press. Rabbi Dayan L. Mendelsohn, M.A., read the proofs and discussed various rabbinical points with me. Captain Maurice A. Ainslie, R.N., Instructor of Astronomy to the Navy, read the book from the Astronomical aspect and Dr. J. K. Fotheringham, Reader of Ancient Astronomy and Chronology at the University of Oxford, did similar service in connection with a portion only of the book. Each of them corrected a number of inaccuracies which could only be detected by expert astronomers. The Rev. Dr. A. Cohen, M.A., read a portion of the proofs and detected several typographical errors. Mr. D. S. Sassoon gave me every facility to examine the manuscripts of *Kiddush Hachodesh* in his possession. Similar facilities were granted to me by the librarians of the Oriental Department of the British Museum, of the Bodleian Library, Oxford, of the Biblioteca Apostolica Vaticana, as well as of the Frankfurt Stadtbibliothek, who placed the manuscripts found in their respective libraries at my disposal, or were good enough to send me photographic reproductions of the passages in which I was particularly interested. My cousin, Mr. Reginald V. Feldman, B.A., translated for me, into English verse, the Hebrew lines which form the heading to Chapter XIV., and the publisher, Mr. M. L. Cailingold, took great pains to detect and correct Hebrew typographical errors. Lastly, it is a very great pleasure to record my grateful appreciation of the honour accorded to my book and myself by such eminent authorities as Professor R. A. Sampson and Professor Adolf Fraenkel, in consenting to introduce the book by means of preface and foreword.

150, Harley St., W. W. M. FELDMAN.
July 1931.

TABLE OF CONTENTS

List of Works Frequently Consulted.

Talmud, with the usual commentaries (in the original). There are English and German translations of the Babylonian text and a French translation of the Palestinian text.

R. Abraham bar Chiyah. Ha-Meschichah V'ha-Tishboreth (המשיחה והתשבורת); Sefer Ha-Ibbur (ספר העיבור).

Maimonides. Hilchoth Kiddush Hachodesh (הלכות קדוש החדש)

Israeli. Yesod Olam (יסוד עולם)

E. Baneth. "Maimuni's Neumondsberechnung."

Ch. E. Bornstein. A Chapter on the History of the Jewish Calendar (in Hebrew) in Sefer Ha-Jobel (Sokoloff's Jubilee Volume, Warsaw 1904).

S. Brodetsky. Astronomy in the Babylonian Talmud, Jewish Review, 1911.

*Sh. B. **Burnaby***. Elements of Jewish and Muhammedan Calendars. London, 1901.

D. Gans. Nechmad V'Naim (נחמד ונעים) Jesnitz, 1743.

F. K. Ginzel. Handbuch der mathematischen und technischen Chronologie. Vol. II. Leipzig, 1911. (Contains up-to-date Bibliography on Chronology.)

E. Hechim. Shwile de Rekia (שבילי דרקיע) Prag, 1785.

E. Mahler. Zur talmudischen Mathematik, Zeitschr. f. Mathematik 1884-1886.
Handbuch der jüdischen Chronologie. Leipzig, 1916. (Contains up-to-date Bibliography on Jewish Chronology.)

H. M. Piniles. Darka Shel Torah (דרכה של תורה) Wien. 1861.

Ad. Schwartz. Der. jüd. Kalender. Breslau, 1872

Ch. Z. Slonimsky. Toldoth Hashamaim (תולדות השמים)
Yesodé Haibbur. (יסודי העיבור).

B. Zuckermann. Das Mathematische in Talmud. Breslau, 1878.

Jewish Encyclopædia.

Ozar Israel (abbreviated Hebrew edition of Jewish Encyclopædia)..

Hastings' Encyclopædia of Religion and Ethics.

PART I.

PROLEGOMENA.

תקופות וגמטריאות פרפראות לחכמה
(אבות, ג׳. י״ח.)

" Astronomy and Mathematics are the Spice of Wisdom."
(Aboth iii, 18)

CHAPTER I.

What is the Talmud?

THE TALMUD (תלמוד) which literally means "the study", is the name given to a work in many volumes, written partly in Hebrew and partly in Aramaic, which embodies the teachings and opinions, on religious and sociological matters, of the ancient Jewish sages, during a period of some 800 years (viz. from 300 B.C. to 500 C.E.). In addition, it contains the theological, theosophical and philosophical views, as well as the moral and ethical maxims of those sages. It is further interwoven with many observations of an anatomical, physiological, medical, anthropological, or scientific character.

The Talmud consists of two distinct portions, viz:

(1) THE MISHNAH (משנה)
(2) THE GEMARA (גמרא)

(1) THE MISHNAH. Authorities are not agreed regarding the exact derivation or meaning of the word Mishnah. According to some it contains the same root as the word *Shnaim* (שנים) meaning "two", and is meant to signify that it is second in rank to the Bible. According to other scholars it is derived from the verb *Shanah* (שנה) which means "to teach orally", and hence it is meant to express the fact that the teaching therein contained was transmitted from generation to generation by word of mouth, in contradistinction of the written law of the Pentateuch, which is called *Mikra* (מקרא) i.e., "that which can be read".

The Mishnah was compiled and edited by the famous Rabbi Jehudah Hanasi (R. Judah the Prince or President of the Synhedrin) known by way of eminence as, simply, Rabbi,—who died about the end of the second century of the present era. It represents the legal traditions (based upon the written law contained in the Pentateuch) of many generations of eminent scholars—called respectively the *Soferim* or Scribes, or men of the great Synod, *Zugoth* or pairs or duumviri,* and *Tanaim* (תנאים) or teachers

* The duumviri were pairs of eminent Scholars who acted as president (Nasi) and Vice-president (Ab Beth Din) respectively of the Synhedrin.

—dating from the time of Simon the Just (310-392 B.C.) to that of Rabbi Judah Hanasi.

The language of the Mishnah is "New" Hebrew, which however is not essentially different from Biblical Hebrew.

DIVISIONS OF THE MISHNAH. The whole of the Mishnah is divided into six main sections called *Sedarim* (סדרים) or orders, and hence the whole of the Talmud is sometimes known as the *Shas* which is a word composed of the first two letters (SH—ש S—ס) of the words *Shisha Sedarim* or "the six sections". These sections are named in accordance with the subject with which they deal, thus:—

Name of Section.	*Subject dealt with.*
Zeraim, or Seeds.	Ritual laws of Agriculture.
Moëd, or Festival.	,, ,, of Sabbaths and Festivals.
Nashim, or Women.	,, ,, of marriage and divorce.
Nezikin, or Damages.	Civil and Criminal Law.
Kódeshim, or Sacred Things.	Temple Services and Sacrifices.
Taharoth, or Purification.	Ritual Laws of Clean and Unclean.

Each section is sub-divided into Masechtoth (מסכתות) or treatises or tractates, of which there are sixty-three in number, each of which is named in accordance with its general contents, e.g., Pesachim deals with the laws relating to Passover and the Paschal lamb. Each Masechta or treatise is again sub-divided into chapters, or *Perakim*, and each chapter consists of a number of paragraphs, each of which consists of a *Mishnah* or *Halachah*. The latter is the name given to each paragraph in the Palestinian Talmud.

(2) THE GEMARA. Whilst there is no difference of opinion regarding the etymology of the word Gemara, as being derived from the word *gamar* (גמר), there is no complete agreement regarding its exact meaning. In Hebrew, *gamar* means "to complete," and hence some authorities believe the word gemara to mean "completion of" or "supplement to" the Mishnah. In Aramaic, however, *gamar* means "to teach" and hence other authorities believe the word gemara to have the same meaning as Talmud, viz: "the teaching", or "the study".

The gemara, which consists of commentaries on the Mishnah, represents the discussions and disputations of the *Amoraim* (אמוראים) or interpreters, who were the directors and members of the famous Babylonian or Palestinian Academies, after the time of R. Judah Hanasi, and who flourished during the three centuries 300-500 C.E.

THE TWO GEMARAS. Each school of Amoraim—the Babylonian and the Palestinian—expounded the Mishnah in somewhat different language and a somewhat different manner from the other, and hence there are two compilations of the Gemara , viz. : -

(i) *The Talmud Babli*—or Babylonian Talmud which was made in Babylon, in the famous Academies of Nehardea, Sura, and Pumbeditha. Its compilation was begun by Rab Ashi, who at the age of twenty was appointed President of the Academy of Sura—an office which he held for 52 years (375-427 C.E.)—and it was completed by his pupils and successors, as well as by R. Josi, the President of the Academy of Pumbeditha.

(ii) *The Talmud Jerushalmi*—or Palestinian Talmud, was compiled in the Palestinian Academies, and inasmuch as there was no Academy in Jerusalem after the destruction of the second temple, the name Jerushalmi is, in a sense, a misnomer.

The two compilations differ in several respects, but the chief difference is one of *scope*. Although the Palestinian gemara covers thirty-nine out of a total of sixty-three Masechtas, or treatises of the Mishnah as against only thirty-seven Masechtas covered by the Babylonian gemara, nevertheless the total material of the former amounts to no more than one third of the latter, and its discussions do not exhibit the same dialectical acumen as is shown in the Babylonian edition. Indeed so acute was the analytical method of study developed in Babylonian Academies that it was jocularly said that in Pumbeditha " they can pass an elephant through the eye of a needle."[1] It is partly on this account and partly also as the result of the fact that the famous commentary of *Rashi* (רש״י)* —which illuminates some of the obscurest passages, and disentangles some of the most knotty arguments of the Talmudical discussions—does not extend to the Palestinian edition, that the study of the latter has never been taken up with any great enthusiasm by any but the most eminent Rabbis.

[1]Baba Metzia 38b. מעיילין פילא בקופא דמחטא

*So called from the initials of his name Rabbi Solomon (*Sh*lomah) *I*tzchaki (son of Isaac) of Troyes (1040-1105).

Note: References to the Babylonian Talmud are given as follows:—

(1) Baba Bathra 101a or 101. means Treatise Baba Bathra, folio 101, side 1.

(2) Succah 7b or 7: means, Treatise Succah, folio 7, side 2. A folio consists of two pages.

(3) Oholoth xii. 6, means Treatise Oholoth, Chapter xii., Mishnah 6.

In the case of those treatises which only contain the Mishnah but no Gemara the system of reference (3) is adopted. In the case of those where the Gemara does exist, the system given under (1) and (2) are used.

Reference to the Palestinian Talmud are distinguished by the addition of the letter J. (standing for Jerushalmi) in front of the name of the treatise. Thus J. Erubin stands for treatise Erubin of the Palestinian Talmud.

CHAPTER II.

Mathematics and Astronomy in Relation to Religious Legislation.

Whilst the primary work of the Talmudic Rabbis lay in expounding and developing the civil and religious law of the Bible, a considerable acquaintance with the various branches of science was necessary in connection with such legislation. The administration of the dietary laws, for instance, demanded not only a classification of all animals into clean and unclean — with the distinguishing signs between them — but it also necessitated a knowledge of morbid anatomy and pathology to be able to recognise, post mortem, whether an animal that was killed for food had previously suffered from some disease which was likely to shorten life—in which case the carcass was unfit for human consumption (*trepha.*) Legislation for the right of the poor to participate in harvests, rules for tithes, and cognate matters, required a knowledge of botany and classification of plants. The determination of the calendar required a sound knowledge of Astronomy, since not only were Jewish festivals fixed on given days of the lunar month, but they also depended on the position of the sun. Passover, for instance, which begins on the 15th day of Nisan must also occur in the month of the wheat harvest or *Abib* (Deut. xvi. 1); Tishri must take place in the autumn. Further, the position of the moon in relation to the sun, as well as its height above the horizon at any moment, had to be computed mathematically, for the purpose of ascertaining whether the new moon could be visible at that given moment. This procedure was necessary for the purpose of fixing the beginning of every new month. Lastly, the laws in connection with the making of an *erub* i.e., an imaginary combination of localities for the purpose of extending the Sabbath boundary—or the distance which the Jew is allowed to walk in a public road, on the Sabbath, from an enclosed private place—and thus circumvent the Biblical injunction of not going out of one's place on the seventh day, as stated in Exodus xvi. 29; legislation regarding the sowing of fields—in which, according to Leviticus xix. 19, two kinds of seed may not be sown within a certain distance of each other; as well as certain other religious observances, demanded a knowledge of Mathematics.

None of these sciences are dealt with in any systematic manner in the Talmud, but scientific problems are alluded to casually, and in the concisest possible manner, in relation only to the theological or religious questions under consideration. Hence we find that the statements of mathematical and astronomical interest are scattered about, in a haphazard way, throughout the many thousands of pages of the numerous treatises of the remarkable compilations comprising the Babylonian and Palestinian Talmuds. These statements either directly express some mathematical fact, (e.g., the relation between the length of the circumference of a circle and its diameter; the relation between the length of the diagonal of a square and its side, etc.); or are such as show that the participants in the discussions must have possessed a considerable amount of skill in mathematical computations and geometrical construction.

That the mathematical or astronomical facts directly or indirectly mentioned in the Talmud, most probably represent a portion only of the total Rabbinical knowledge of this subject, can be inferred from the fact that great reticence was observed with regard to matters of astronomical theory. Indeed, so jealously was the secret of the calendar (סוד העיבור) guarded that R. Chanina, the Palestinian who constructed a calendar outside Palestine, was reprimanded for committing such a reprehensible act, by R. Josi ben Kepar and another, who were specially sent from Jerusalem for that purpose. His plea that R. Akiba did a similar thing in Nehardea was met by the rebuke that whilst the latter left no one in Palestine equal to the task of calendar making, the same was not the case with R. Chanina, for " the kids you left in Palestine have now become horned goats " (גדיים שהנחת נעשו תישים בעלי קרנים) " and it is they that sent us to you."[1]

[1] Berachoth 63a. Compare Pirké d' R. Eliezer viii. אפילו צדיקים חכמים בחוצה לארץ ורועי צאן ובקר בארץ, אין מעברין את השנה אלא על ידי רועי צאן ובקר.

Such secrecy regarding astronomical matters was not peculiar to Jews alone, but was the prevailing custom among all ancient astronomers who spoke in riddles with the object of mystifying the public and thus lend special aggrandisement to this Science. (See Isaac Israeli, Yesod Olam, ii. 6).

Interpretation of Mathematical Discussions in the Talmud.

To get the sense of any Talmudic discussion is a matter of extraordinary difficulty, and requires years of intensive study and experience. This is so for the following reasons:-

(1) The language difficulty. The text being chiefly Aramaic and the idiom being so peculiar, even the best scholar of Biblical Hebrew who has not received any training in the Talmud cannot understand it.

(2) The great verbal economy of the Rabbis. The sentences are so concise as to leave any tyro merely perplexed, and entirely at a loss to interpret their meaning.

(3) Absence of punctuation. All sentences are jumbled together without commas, signs of interrogation, or even full stops, and considerable experience is needed to separate one sentence from another, or to tell where a question ends and an answer begins. When we come to mathematical discussions, the difficulties are still further increased by—

(4) The absence of diagrams. Practically all the mathematical problems discussed are of a geometrical character, but the text contains no diagrams to help the student in their elucidation. Indeed, were it not for the wonderful running commentary of Rashi, who sheds floods of light upon the most cryptic passages, the Talmud would undoubtedly have remained a closed book to the largest majority of students. In the mathematical portions he adds diagrams which are of some help to the bewildered student.

One cannot speak of this commentary without contemplating with deep and reverent admiration the most wonderful genius and extraordinary versatility of its author. For, as I shall show later, Rashi was no mathematician and yet he was able to work out each mathematical problem, as he met it, by a logical application of first principles (see p. 58).

There are numerous other commentaries of which the chief is the Tosefoth (meaning addenda). This is not a mere elucidation of the text, as in the case of Rashi, but is of a critical character. It seeks out apparent contradictions between one Talmudic passage and another, and in a most ingenious manner attempts to harmonise them. This commentary is the work of many authors of the 12th and 13th centuries, amongst whom are two sons-in-law and one grandson of Rashi.

Maimonides (12th century) also wrote a commentary on the whole of the Mishnah, (begun at the age of 23 in Spain and finished at the age of 30 in Egypt). His *Hilchoth Kiddush Hachodesh* which deals with the astronomy of the calendar was written in 1178, when he was 38 years old.

All these commentaries will be considered as occasion arises.

CHAPTER III.

Some Mathematical Celebrities of the Talmud.

R. Yochanan ben Zaccai a distinguished teacher of the first century, C.E., and founder as well as Principal of the celebrated Academy at Yabne—was the youngest disciple of the great Hillel and occupied himself amongst other things with the study of astronomy and mathematics (תקופות וגמטריאות)[1]

R. Gamliel the Elder (1st century C.E.) — a grandson of Hillel, whom he succeeded in the office of Nasi or President of the Synhedrin, was a very brilliant mathematical astronomer. The Mishnah states that he had charts illustrating various phases of the moon's crescent, hanging in his observatory (בעלייתו) and he used these for testing the evidence of the laymen (ההדיוטות) who came to report regarding the visibility of the New Moon[2]. It may be stated en passant, that such determination of the New Moon by actual observation is said to have been carried on till the beginning of the Middle Ages.

According to R. Ishmael[3] there were 3 things which Moses found it difficult to understand, until God pointed them out to him with His finger; one of them was the shape of the lunar crescent at New Moon.

The following episode related in the Mishnah [4] illustrates the great confidence that R. Gamliel had in his astronomical calculations: "Once two witnesses reported that they saw the moon in the East *before* sunrise and in the West *after* sunset, when R. Yochanan ben Guri considered their evidence unreliable. When they came to Yabne, R. Gamliel accepted their evidence, but R. Dosa ben Hyrcanus insisted that their evidence was false; for, on R. Nachman's authority,[5] the moon cannot be seen within an interval of 24 hours both *before* sunrise and *after* sunset and therefore, as he picturesquely, though somewhat crudely, contends ' how

[1] Succah 28a [2] Rosh Hashana 24a [3] Menachoth 29a

[4] Rosh Hashana, ii. 8. (24b and 25a)

[5] Ibid. 20b. כ"ד שע' מכסי' סיהרא

is it possible to testify about a woman having given birth on one day when the next day her stomach is between her teeth '?"

(היאך מעידים על האשה שילדה ולמחר כריסה בין שיניה)

i.e., how is it possible for the old moon to die and for the new moon to have appeared within so short an interval of time?

R. Joshua (R. Gamliel's Junior Colleague) agreed with R. Dosa. But such was R. Gamliel's perfect faith in his own calculation and, therefore, in the incorrectness of that of R. Joshua, that he ordered the latter deliberately to desecrate the day which according to R. Joshua's reckoning should be the Day of Atonement (10th of Tishri), by coming to him on that day with his staff in his hand and money in his pocket. R. Joshua was advised by R. Akiba that R. Gamliel's authority was supreme, and he did as he was bidden. R. Gamliel then embraced R. Joshua and said to him: 'Welcome, my teacher and my pupil! My teacher art thou because thou art so wise, and my pupil because thou hast obeyed my commands.'"

The point in this incident is that —as we shall see in a subsequent chapter— the moon is *East* of the sun throughout the first half of the month, and *West* of it during the second half. Hence the *New Moon* cannot be seen in the morning at the eastern horizon until *after Sunrise*, and in the evening at the western horizon until *after Sunset*; whilst the *Waning Moon* can only be seen in the morning *before Sunrise* and in the evening *before Sunset*. If, therefore, witnesses testify to having seen the moon before sunrise and after sunset their evidence cannot be true—unless it refers to the waning moon in the morning and the new moon in the evening. But as there must be a certain minimum angular distance between the sun and moon for the latter to be visible, therefore between the morning —when the moon must have been that minimum angular distance *behind* (or to the west of) the sun—and the evening—when it must have got that minimum angular distance *ahead of* (or to the east of) the sun, the moon must have covered double that angular distance. But, at the rates at which the sun and moon are travelling, this would be impossible. Hence R. Yochanan ben Guri rejected their evidence. R. Gamliel, however, who by his astronomical calculations concluded that the new moon was indeed visible on the evening in question, accepted their evidence, on the assumption that what they believed to have seen in the morning was a piece of cloud that looked like the very thin crescent of the waning moon.

In subsequent pages we shall show how R. Gamliel probably worked out the problem of the moon's visibility. (See Chapter XV).

R. Gamliel is also said to have possessed an instrument resembling a telescope or astrolabe (שפופרת = tube) by means of which he could estimate a distance of 2,000 cubits, as well as measure depths and heights.[1] (See p. 40.)

The same *R. Joshua ben Chananya* was a disciple of R. Yochanan ben Zaccai. He founded an Academy at Bekiin and was also a member of the Synhedrin at Yabne—of which R. Gamliel was president or Nasi. The following story is narrated about him and R. Gamliel.[2] " R. Gamliel and R. Joshua took a sea voyage together. R. Gamliel had with him a supply of bread, whilst R. Joshua had a supply of bread as well as of fine flour. After R. Gamliel had eaten his bread, he had to use R. Joshua's flour. Said R. Gamliel to R. Joshua: 'Did you know that the voyage would last so long that you provided yourself with flour? And R. Joshua replied: 'A star appears every 70 years and leads ships astray (by causing the captains to steer by it erroneously), and I thought to myself: it might appear and lead us astray.' Then R. Gamliel said to him in surprise: 'You are such a clever man and yet you are so poor as to have to travel abroad to make a living?' And R. Joshua replied bitterly: 'Are you surprised at me? Rather be surprised at your pupils, R. Eliezer Chisma, and R. Yochanan ben Gudgada, who can estimate the number of drops in the ocean and yet have neither bread to eat nor clothes to wear. " This was a rebuke to R. Gamliel for having refused to promote such eminent scholars as these two pupils of his were.

The above story obviously relates to some comet which appears every 70 years, and it has therefore been suggested[3] that R. Joshua was acquainted with Halley's comet (whose periodic time is about 75 years). According to Brodetsky,[4] however, this view is untenable, since one of the periodic returns of Halley's comet was in the year 66. But the voyage referred to could not have taken place in that year, since R. Gamliel was at that time a minor and could not

[1] Erubin 43b. [2] Horyoth 10a.

[3] See letter by J. L. Rapoport in Slonimsky's Toldoth Hashamayim (תולדות השמים) 1st edition, 1838.

[4] "Astronomy in the Babylonian Talmud." *Jewish Review*, May 1911, p. 72.

have had pupils. Also, as R. Gamliel died about the year 115 C.E, the voyage could not have taken place after that year. His journey to Rome took place about the year 90.

Mar Samuel (165-250 C.E.), a pupil of R. Judah Hanasi, was a very great Rabbi and Physician, as well as an eminent astronomer, and was therefore called *Yarchinai* (which means 'Lunar Expert' or 'the Astronomer'). His proud boast was that he was "as perfectly familiar with the roads of the skies, as he was with those of his own native town of Nehardea. Nevertheless," continued he, in a spirit of humility, possibly to counterbalance his apparent arrogance, "I am ignorant of the nature of '*Cochba de-Shebit*.'"[1] There is some difference of opinion as to what Cochba de-Shebit means. 'Cochba' means a star, and Cochba de-Shebit is held by Rashi to mean a shooting star, whilst others, including Slonimsky, interpret it as a comet. Selig Brodetsky[2] is inclined to agree with Rashi.

The same Samuel was once rebuked by another Rabbi for boasting about his great astronomical skill.[3] "I can fix the calendar," said Samuel, "for the whole of the diaspora, when Abba the father of R. Simlai said to him, "Do you know the meaning of the statement in connection with the secret of Intercalation which distinguishes between the new moon before 12 o'clock and that after 12 o'clock?" When Samuel acknowledged his ignorance in this respect, Abba remarked caustically, "If you do not know this, there may be other things of which you are equally ignorant." (See p. 192). The famous Greek astronomer and mathematician Thales (640-546 B.C.) who fell into a ditch whilst contemplating the heavens was similarly reproved by a woman who said to him: "How can you see what is going on in the heavens if you cannot see what is at your feet?"; and R. Gamliel is recorded as having used almost identical words to reprimand a man who, though ignorant of the number of his teeth, presumed to be able to know the number of stars.[4]

On another occasion Samuel, apparently in a spirit of boastfulness, sent R. Yochanan a list of leap years for 60 years, and the latter cynically replied that any mediocre mathematician (חושבנא בעלמא) can do that.[5]

[1] Berachoth 58b. [2] Loc. cit July 1911, p. 168. [3] Rosh Hashanah 20b.
[4] Synhedrin 39a. דאיכא בפומך לא ידעת דאיכא ברקיע ידעת
[5] Chulin 95b.

Maimonides in his commentary to Rosh Hashanah (ii. 4) utters a similar rebuke to those who, in virtue of their ability to calculate the calendar (according to the method of the " Fixed Calendar " q.v.), lay claim to mathematical skill which they do not possess:

אבל כולם [בעלי העבור] אין בידם אלא מלאכת הקיבוץ לבד ר"ל קיבוץ מנין למנין אחר וידמה שהוא התגבר על חכמה גדולה, וכו'.

For as Maimonides points out in his *Hilchoth Kiddush Hachodesh* xi. 4. " the method of the fixed calendar is one which an average school child can master in 3 or 4 days ".

זה החשבון.... בזמן הזה שאנו מחשבין בו היום אפילו תינוקות של בית רבן מגיעין עד סופו בשלשה וארבעה ימים.

R. Eliezer Chisma and *R. Yochanan ben Gudgada* were said to be able to count the drops in the ocean, " and yet," adds R. Joshua ben Chananyah bitterly, " they had neither bread to eat nor clothes to wear."[1] After R. Joshua's rebuke (see p. 11) R. Gamliel sent for them' offering them promotion, but in their great humility they refused to accept it. He then sent for them a second time and said : " Do you suppose I am offering you greatness? It is slavery that I am offering you since it is said (I Kings xii, 7) that a king is a servant of his people." R. Eliezer Chisma is responsible for the saying: " Astronomy and Geometry are the crowning glory of wisdom." [2]

Other Rabbinical astronomers and mathematicians were R. *Eliezer the Great* and *R. Josi b. Chalafta* — generally known as R. Josi — both of whom flourished in the 2nd century, and were disciples of R. Akiba. R. Josi clearly defines the different positions of the sun in the heavens at different times of the year as well as the relation between the ecliptic and the equator (see under " Meridian ").[3] *Rabbi Nathan* of the 2nd century was also well versed in mathematics and astronomy. *Abaye* of the 3rd century understood the motions of the planets very well and knew the 28 years' Solar Cycle (q.v.).[4]

1 Horyoth 10a. Compare the statement of Archimedes (3rd century B.C.) that he could count the sand in the universe. On the assumption that 1000 grains of sand occupied the same volume as a poppy-seed of diameter $\frac{1}{40}$ finger-breadth, he calculated that a heap of sand as large as the whole Universe (i.e. a sphere of radius equal to the distance between the earth and the stars) would contain 10^{63} grains of sand.

2 Aboth iii. 23. 3 Erubim 56a. 4 Berachoth 59b.

PART II.

MATHEMATICS.

ת"ר לא תעשו עול במשפט במדה במשקל ובמשורה. במדה זו מדידת קרקע שלא ימדוד לאחר
בימות החמה ולאחר בימות הגשמים (ב"מ סא:)

The Rabbis taught as follows: The verse (Lev. xix. 15), 'Ye shall do no unrighteousness in judgment' applies to mensuration of land, as well as to the weighing and measuring of solids and fluids, so that one should not use the same measuring rope in Summer and Winter (on account of the errors arising from differences in hygrometric conditions). Baba Metzia 61b.

CHAPTER IV.

Arithmetical Numeration and Notation of the Rabbis.

Whilst in the Bible the system of Numeration, whereby numbers are expressed in words, is constantly employed, the Talmudic Rabbis used a system of Notation, i.e., a method of expressing numbers by means of symbols, and, like the Greeks, utilized the letters of the Alphabet for that purpose; each letter up to the 10th having its place value. Thus א = 1; ב = 2; ג = 3; and so on up to י = 10; יא = 11, etc., up to יט = 19. The 11th to the 19th letters represent multiples of 10 up to 100. Thus the letter כ = 20; ל = 30; מ = 40; נ = 50; ס = 60; ע = 70; פ = 80; צ = 90; ק = 100. The last letters of the Alphabet represent multiples of 100, ר = 200; ש = 300; ת = 400.

The Rabbis were also aware of the Decimal System of Notation, which, they state, they adopted from the Persians.[1]

Gematria.

The alphabetical system of notation was utilised a great deal in the Talmud for the purpose of making certain mnemonics or deductions, which depended upon the numerical equality of words. This method is called *Gematria*.[2] The following are a few examples :-

(1) The duration of pregnancy was taken as 271 days because the numerical value of הריון which means pregnancy is $5 + 200 + 10 + 6 + 50 = 271$ [3]

(2) " When wine enters, the secret escapes "[4] because יין (wine) and סוד (secret) have the same numerical value of 70. (יין ניתן בע' אותיות וסוד ניתן בע' אותיות נכנס יין יצא סוד)

(3) The Patriarch was first called Abram (אברם) because God

[1] Bechoroth 60a. (בכ:ינא פרסאה דקרו לעשרה חד)

[2] The word Gematria has two meanings and probably two derivations, in the Talmud. One is the sense in which it is used here, when the word is probable derived from γραμμτεια which is a derivative of the word γραμμα = word. The other meaning is mathematics, or calculation in which case it is derived from γεομετρια = calculation.

[3] Niddah 38b. [4] Erubin 65a.

> originally gave him command over the 243 bones in his body (אברם = 243). Later he was named Abraham (אברהם) because he was also made master over his two eyes, two ears and his genital organ (ה = 5) making altogether 248 parts[1]

The term *digit* for the symbols representing numbers is, of course, derived from the Latin *digitus* meaning finger, since counting was in ancient times done on the fingers. Indeed the decimal system originated from the fact that we have 10 fingers. Such method of counting is referred to in the Talmud.[2] Numbers were often formed by subtraction. Thus $666\frac{2}{3}$ is referred to as 667 less $\frac{1}{3}$.[3] This method was used by the Romans and is still adopted by the French.

[1] Nedarim 32b. [2] Yoma ii, 1 & 22b. (הצביעו... תנא הוציאו אצבעותיכם למנין)
[3] Erubin 57a.

CHAPTER V.

Mathematical Data and Theorems.

(A) ORDINARY ARITHMETICAL PROCESSES.

The four basic arithmetical rules (addition, subtraction, etc.) were used for integers as well as fractions.

(a) *Reduction of Measures.*

" R. Chisda said that the volume of a Rebiith (רביעית) is that of a prism with a square base 2 by 2 fingers' breadths and of a height two and a half, and a fifth, fingers breadths."[1]

This of course means that a Rebiith $=4(2\frac{1}{2}+\frac{1}{5})=4\times 2{\cdot}7=10{\cdot}8$ cubic fingers' breadths.

He proves his statement as follows: " The size of a ritual bath or Mikvah (מקוה) is 3 cubic cubits and this contains 40 Seah of water."

The actual calculation is not given, but we know that 1 Seah $=$ 6 Kab; 1 Kab $=4$ Log; and 1 Log $=4$ Rebiith.

Therefore 1 Seah (סאה) $=6\times 4\times 4=96$ Rebiith.

Therefore 40 Seah $=96\times 40=3840$ Rebiith.

Again we know that

1 cubit $=6$ Handbreadths, and 1 Handbreadth $=4$ fingerbreadths.

Therefore 1 cubit $=6\times 4=24$ fingerbreadths.

Therefore 1 cubic cubit $=24^3=13824$ fingerbreadths.

Therefore capacity of Mikvah which $=3$ cubic cubits $=41472$ fingerbreadths.

But capacity of Mikvah $=40$ Seah $=3$ cubic cubits.

Therefore 3840 Rebiith $=41472$ fingerbreadths.

Therefore 1 Rebiith $=\frac{41472}{3840}=10{\cdot}8$ cubic fingerbreadths

[1] Pesachim 109 a & b.

(β) *Problem in Proportions.*

The following is an interesting arithmetical conundrum: " R. Pappa and R. Huna the son of R. Joshua once dined together and R. Pappa ate 4 times as much as R. Huna . . . . On another occasion R. Huna and Rabina dined together and Rabina ate 8 times as much as R. Huna. Then, said R. Huna: I would rather dine with 100 people like R. Pappa than with one Rabina."[1]

This apparent exaggeration is a perfectly correct statement from the arithmetical standpoint. What it means is that if he had to pay in proportion to the number of people that were dining, then, when eating together with Rabina, the amount of food consumed between them would be 9 times as much as R. Huna ate, and he would therefore have to pay for $4\frac{1}{2}$ portions.

When dining with 100 people like R. Pappa the amount of food consumed between them would be 401 times as much as R. Huna took, and as there were 101 diners he would have to pay for $\frac{401}{101}$ portions, which is of course less than for $4\frac{1}{2}$ portions.

(γ) *Test for the divisibility of a number by 7.*

The Sabbatical year is the name given to every 7th year during which the ancient Hebrews were not to sow their fields, or prune their vineyards. (Exod. xxiii, 10, 11; Lev. xxv, 2-7; Deut xvi, 1-11 and xxxi 10-13). In order to be able to determine whether any given year was a Sabbatical year the following test is given: " Multiply the hundreds in the given number by 2, and add the product to the tens and units of that number. If that sum is divisible by 7, the whole number is divisible by 7."[2]

This rule, the reason for which is not given, of course depends upon the fact that $100=98+2=(7\times 14)+2$. Hence such a number as $224=2(98+2)+24$. Now 2×98 is divisible by 7. Therefore as $2\times 2+24$ which$=28$ is also divisible by 7, the whole number must be divisible by 7.

(B) GEOMETRY.

The treatises which devote a good deal of space to the consideration of questions involving a knowledge of practical geometry and of astronomy are the following:-

[1] Pesachim 89b. [2] Aboda Zarah 9b.

(1) Erubin, which treats of the distances that one may walk on the Sabbath day. This, together with the treatises Sabbath, Pesachim, Succah, Rosh Hashanah, and seven others, belongs to the section Moed.

(2) Succah, deals with the laws concerning the feast of Tabernacles.

(3) Rosh Hashanah, considers problems connected with the feast of the New Year and kindred astronomical problems.

(4) Synhedrin deals with subjects relating to the Calendar.

(5) Kilayim, which concerns itself with the problems of the prohibited mingling of seeds of cattle and of plants as well as of the mixing of linen and wool in garments (Lev. xix, 19; Deut. xxii, 9-11). This treatise belongs to Section Zeraim, which contains 11 treatises altogether.

(6) Mikvaoth, which discusses the laws concerning baths which are to be used for ritual purification. It belongs to Section Taharoth, which contains altogether 12 treatises.

In addition, the treatise Middoth gives the measurements of the Temple, its courts, etc.

A few other treatises also contain some mathematical problems. These will be mentioned in the course of this work.

NOTE: In the translation of all the quotations given hereafter, the words in brackets have been inserted by the present author, in order to make the meaning clear. In most cases they represent the explanations given by one or other of the commentators mentioned before (see Chapter II.)

GEOMETRICAL CONSTANTS.

1. *The size of the diagonal of a square* is given as equal to 1·4 times the side of the square. The exact statement is as follows: "Every cubit in (the side of) a square, (corresponds to) one and two fifths of a cubit in its diagonal."[1] (כל אמתא ברבוע, אמתא ותרי חומשא באלכסונה).

This statement occurs a number of times in connection with various discussions of practical problems (see next chapter).

The Tosefoth Commentary (12th century) points out that this value falls somewhat short of the real value and proves it in the following simple manner:-

[1] Erubin 57a and other places.

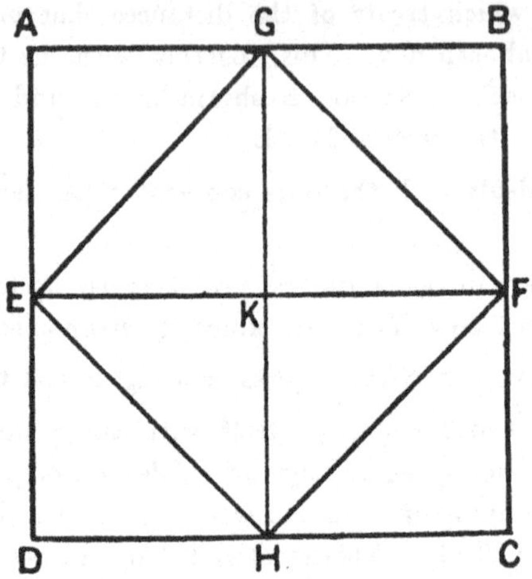

FIG. 1.

Let side of square ABCD (Fig. 1) = 10 units, then area of square ABCD = 100 sq. units. Bisect the sides at GHFE, and join GF, GE, FH, HE, Then since AG and AE are each 5 units, therefore if the ratio required $= 1.4$ then $EG = 5 \times 1.4 = 7$ units.

Therefore the square EGFH = 49 sq. units.

But EGFH being equal to half the area of ABCD = 50 sq. units.

Therefore EG, GF, etc., are each a little greater than 7.

Therefore Ratio between diagonal and side is greater than 1.4.

Compare also Erubin 60b, 78a, 51a, Succah 8a, Sabbath 85a, Baba Bathra 102a, Oholoth xii, Kilayim V. (See further under Surds.)

R. Abraham bar Chiya (1065-1136) gives the value of this ratio as approximately $1\frac{2}{5} + \frac{1}{70}$, i.e., 1·4142 which is correct to 4 decimal places.

2. *The value of π* (יחס קוטר העגול אל הקיפו)

The following statement is frequently made in the Talmud: " The circumference of a circle is 3 times as long as its diameter."[1] (כל שיש בהקיפו ג' טפחים יש בו רוחב טפח)

In view of the fact that the Talmud was completed as late as the 5th century C.E., and that Archimedes, in the the 3rd century B.C. as well as the Egyptians, had given more exact values, it is somewhat astonishing that the Talmudic rabbis should have been satis-

[1] Erubin 13b & 14a and other places.

fied with 3 as the value of this ratio. But, as Zuckermann[1] points out, it would seem that the rabbis were aware of more exact values, but accepted the value of 3 as a workable number for religious purposes, since an anonymous rabbi asks on what grounds this value is taken for the ratio. It is in answer to this question (מנא הני מילי) that R. Yochanan gives a Biblical authority for the value of this ratio, by stating that the verse 1 Kings, vii, 23, says that the 'Sea of Solomon' was circular and measured "ten cubits from brim to brim . . . . and a line thirty cubits did compass it round about." In connection with no other mathematical statement is there any similar doubt raised. Indeed as early as the 2nd century C.E. the 49th Middoth of R. Nathan[2] gives the value of π as $\frac{22}{7}$

R. Abraham bar Chiya gives its value as a little less than $3\frac{1}{7}$. Indeed, by taking the more accurate value of π as 3·141593 it is less than $3\frac{1}{7}$ or 3·142857 by only 0·001264.

The Biblical value of $\pi=3$, was used by the Babylonians. The Egyptians assumed its value to be $3\frac{13}{81}$ or 3.16, since they give the area of a circle (i.e. πr^2) as the square of $\frac{8}{9}$ of the diameter, i.e. $\frac{64}{81}d^2$ or $\frac{256}{81}r^2$. Therefore $\pi=\frac{256}{81}=3.16$.

Archimedes (3rd century C.E.) gives the value of π as lying between $3\frac{10}{71}$ and $3\frac{10}{70}$. The early Chinese gave it the value of 3, as well as of $\sqrt{10}$ (which of course is similar to that of the Egyptians, viz., 3.16). In the 5th century C.E. Tsu Ch'ung Chih[3] gave π a value of between 3.1415926 and 3.1415927.

It is interesting to note that the Biblical statement quoted above shows that the value of this ratio was found by actual measurement rather than by logical mathematical reasoning.

Maimonides (11th century) says that it is impossible to calculate this ratio exactly (see p. 59).

3. *Relation between the Circumference or area of a circle and the Perimeter or Area of its escribed square.*

The Mishnah states[4] that "(the perimeter of) a square is (by)

[1] "Das Mathematische im Talmud". Breslau 1878 p. 23.
[2] See Article "Mathematics" in the Jewish Encyclopædia.
[3] See Florian Cajori A History of Mathematics." New York, 1919.
[4] Oholoth xii 6, Succah 8a etc.

a quarter (of itself) larger than the (circumference of its inscribed) circle. (מרובע יותר על העיגול רביע)

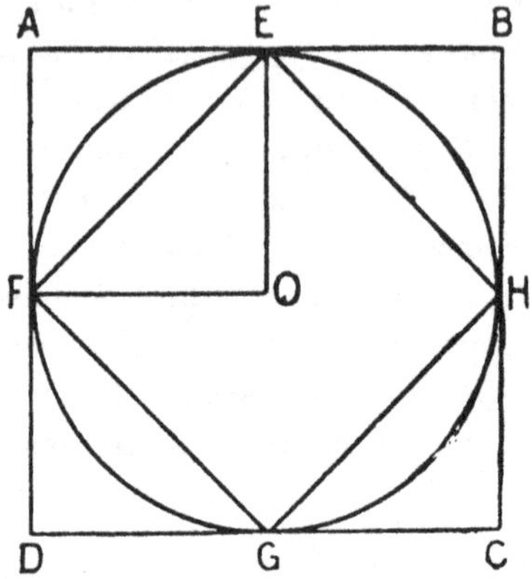

FIG. 2.

In other words the circumference of a circle is $\frac{3}{4}$ of the perimeter of its escribed square. This of course is obvious from Fig. 2, if we take $\pi = 3$, since the perimeter of the square $= 4AB = 4$ times the diagonal, whilst the circumference $= 3$ times the diagonal.

It is interesting to note that a certain R. Chanilai[1] mistook the Mishnah's statement to mean also that the diameter of a circle is $\frac{3}{4}$ of the diagonal of its escribed square, thus making the diagonal equal to $\frac{4}{3}$ of the side of the square. In this he was corrected by R. Ashi (see p. 40).

The same relation would hold good for the areas instead of the lengths as seen in the next theorem.

4. *The relation between the area of a circle and that of its inscribed and escribed squares.*

The Talmud states: "The circle inscribed in a square is a quarter" (i.e. is a quarter less than the area of the square); "the square inscribed in a circle is a half" (i.e. half the area of the escribed square). עיגולא מגו ריבועא ריבעא, ריבועא מגו עיגולא פלגא[2]
The words in brackets are the explanation given by Tosefoth.

This of course is evident from the diagram (Fig. 2) if accept 3 as the value of π.

$$\text{Area of circle} = \pi r^2 = 3FO^2 = \tfrac{3}{4}BC^2 \quad (\text{since } FO = \tfrac{1}{2}BC)$$

[1] Erubin 57a. [2] Ibid. 76b and Succah 8a, b.

Area of $ABCD = BC^2$, $\therefore$ Area of circle $= \frac{3}{4}$ area of ABCD...(i)
Also area of $EFGH = FE^2 = 2FO^2 = 2AFOE = \frac{1}{2}ABCD$...(ii)

From (i) we see that the area of a circle is acknowledged to be $= \pi r^2$

It is interesting to note that these Theorems are quoted in the Talmud "on the authority of the sages of Caesarea" (דייני דקיסרי ואמרי לה רבנן דקיסרי אמרו) which at that time was a Roman province.[1]

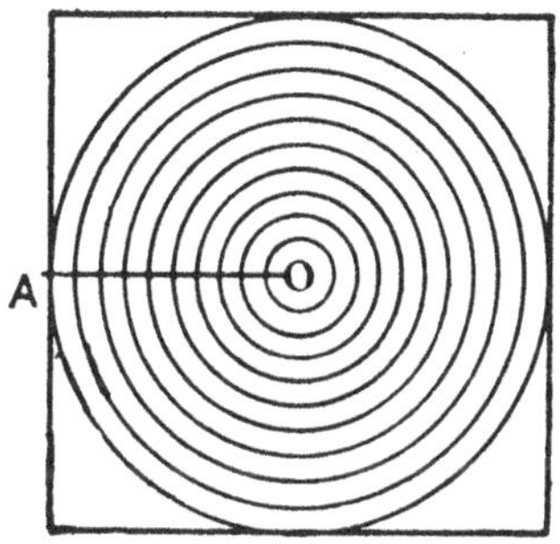

FIG. 3.

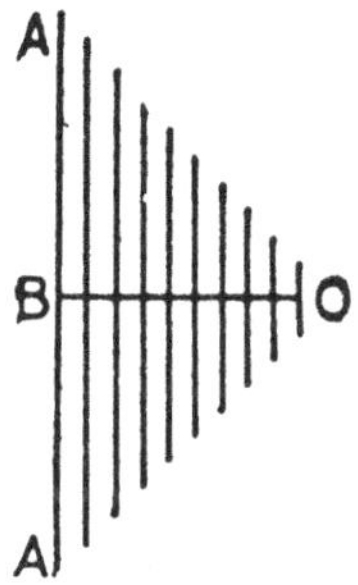

FIG. 4.

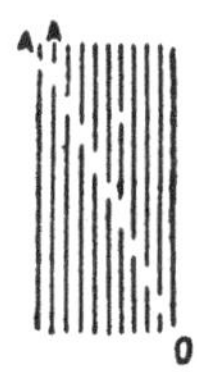

FIG. 5.

Tosefoth proves that the area of a square is $1\frac{1}{3}$ of that of its inscribed circle as follows: Imagine a series of concentric rings, the largest of which has a radius of $\frac{1}{2}$, inside a square 1 by 1. Cut these rings in the direction AO (Fig. 3) and straighten these rings out as in Fig. 4. You then get a triangle whose base is the length of the circumference of the largest circle, viz. $2\pi.\frac{1}{2} = \pi = 3$

Now divide the triangle in the direction of the perpendicular OB and place the two halves side by side as shown in Fig. 5. You then get a rectangle whose side are $\frac{1}{2}$ and $1\frac{1}{2}$ respectively, whose area, therefore, $=3$ times that of a square $\frac{1}{2}$ by $\frac{1}{2}$. But a square 1 by 1 contains 4 squares of $\frac{1}{2}$ by $\frac{1}{2}$.

Therefore area of the square $= \frac{4}{3}$ that of its inscribed circle.

This proof is analogous to that given by Archimedes.[2]

R. Abraham bar Chiya[3] deplores the fact that French surveyors were in his time (early 12th century) ignorant of the elementary

[1] Caesaria, the Capital of Cappadocia (in Asia Minor) belonged to Persia from 530 to 330 B.C. when it was taken by Alexander the Great. Herod captured it in 17 C E.

[2] See Florian Cajori, op. cit.

[3] Preface to his Text Book on Geometry & Mensuration.
(חבור המשיחה והתשבורת, הקדמת המחבר)

principles of mensuration so that when they had to divide land amongst heirs or partners they often allotted ¼ to one who was entitled to ⅓ and vice versa. The same reproach cannot (says he) apply to our Talmudic Rabbis, for where they gave very inaccurate values to certain geometrical constants—such as the size of the diagonal of a square, or the circumference of a circle, or the relations between squares and their inscribed or circumscribed circles —these only applied to ritual measurements, such as Sabbath distances, and even then the effect would be to tighten rather than loosen the law, but when mensuration had to be done for civil purposes the Rabbis were particular that the surveyors carried out their tasks with meticulous care. For the Talmud[1] says that "the verse 'Ye shall do no unrighteousness in judgment' (Lev. xix. 15) relates, among other things, to the surveying of land for civil purposes, so that the surveyor should not use the same measuring line in the heat of the summer (when owing to its expansion the share allotted will be more than the person is entitled to get) as in the winter (when owing to contraction the owner will get less than his due share)"

This explanation of R. Abraham bar Chiya is ingenious but not satisfactory, for it is inconceivable that once they gave rough values for ritual purposes they would refrain from giving more accurate values for civil purposes—if they were in possession of such knowledge.

5. *Relation between a^2 and $(2a)^2$.*

The Palestine Talmud[2] relates the following story, which illustrates how a knowledge of the fact that the area of a square whose side is **2a** is equal to 4 *times* that of side **a**, was utilised to defeat the extortionate demands of an ignorant but despotic Roman Emperor. The Emperor once decreed that the Jews should bring him enough wheat to cover a space of 40 by 40 cubits. They consulted R. Huna who advised them to ask the king to allow them to bring it in two instalments, each time enough to cover an area of 20 by 20 cubits. They obtained this permission and in this way paid only half the amount demanded.

A similar anecdote is related in the case of 2 Arabs some 500 years later. One bought a piece of land 100 by 100 cubits from

[1] Baba Metzia 61b. [2] J. Sotah 20a.

the other, and was offered by the vendor 2 pieces of land 50 by 50 cubits instead, which the purchaser refused. The case was brought before one judge, who, being ignorant of mathematics, decided in favour of the vendor. When the purchaser appealed before a more enlightened judge he won his case.[1]

[1] Dietrici "Propeudeutik der Araber in Xten Jahrhundert", p. 35 (quoted by Zuckermann op. cit. p. 43).

CHAPTER VI.

Talmudic discussions in which the various Theorems are utilised.

1. Size of a circular booth, fit to be used on the feast of Tabernacles. (סוכה העשויה ככבשן).

R. Yochanan said: " A booth (or tabernacle) built in the form of a kiln (i.e., circular) having a circumference sufficiently long to seat 24 persons, is fit for use; if not it is unfit." The Talmud then begins the discussion as follows: " The statement no doubt relies upon Rab's ruling that 'a booth less than 4 by 4 cubits is unfit'. If so, then in view of the facts that (i) one person occupies one cubit, and (ii) that the circumference is 3 times the diameter, a circumference sufficient to accommodate 12 people should be sufficient." [1]

This of course is obvious from Fig. 2, where ABCD is a booth 4 by 4 cubits, which is therefore fit for use. Hence the booth built in the form of the inscribed circle should also be fit; and such a circle has a circumference of only 12 cubits.

Then an anonymous rabbi answers: " Your contention is true with regard to a circular booth, but Rab's ruling applies to a square booth (of 4 by 4), whose perimeter is of course more than 12 cubits."

The questioner is unsatisfied and pursues his cross-examination: " Let us see, the circumference is $\frac{3}{4}$ of the perimeter of the escribed square and therefore its periphery is 16 cubits, so that acccommodation for 16 people instead of 24 should be sufficient?"

The rabbi then answers: " Your contention is true again, but you are assuming that the booth is a circle which is inscribed in a square 4 by 4, whilst R. Yochanan's statement applies to one which is circumscribed about such a square; and such a circumference would be more than 16."

Again the questioner objects: " We know that the diagonal of a square is equal to $1\frac{2}{5}$ of its side, and therefore when the side is 4 cubits the diagonal is 4 times $1\frac{2}{5} = 5\frac{3}{5}$ cubits. But this diagonal forms the diameter of the circumscribed circular booth (see Fig. 2) and therefore the circumference of the latter should be 3 times $5\frac{3}{5} = 16\frac{4}{5}$ cubits or as the Talmud puts it 17 less $\frac{1}{5}$ (instead of the 24 stated)?"

[1] Succah 7b, 8a.

Then the rabbi replies that R. Yochanan was not quite particular about the exact size of the circumference " (ולא דק)

But, pursues the relentless cross-examiner, surely you could not say that R. Yochanan would ignore a difference between $16\frac{4}{5}$ and 24? Then Mar Keshisha, the son of R. Chisda, takes up the matter and says to Rab Ashi, " Your premises are wrong. One person does not take up as much as one cubit, but 3 persons take up 2 cubits." Still the persistent cross-examiner was not satisfied: " Granted that this is so, then 24 persons need only 16 cubits, but the circumference of our booth would contain $16\frac{4}{5}$ cubits?"

And the answer to this is that R. Yochanan was " not particular " about such a trifling difference.

This answer fails to satisfy Rab Ashi, who protests that R. Yochanan might possibly not be particular about such a difference if the difference were the other way about—viz., if he were to require the booth to have a larger circumference than the minimum as required by the law (i.e., more than $16\frac{4}{5}$ cubits), but not to allow a smaller circumference (viz., 16 cubits), than the law demands.

Apparently Mar Keshisha cannot answer this and R. Assi offers the suggestion that R. Yochanan really considered that each person occupies 1 cubit, and therefore 24 persons need a circumference of 24 cubits. Now the diameter of such a booth would, of course, be 8 cubits. But R. Yochanan meant that the circumference of the booth must be such that 24 people could encircle it from the outside rather than the inside (or as the Talmud puts it " R. Yochanan does not reckon the space occupied by the person ").

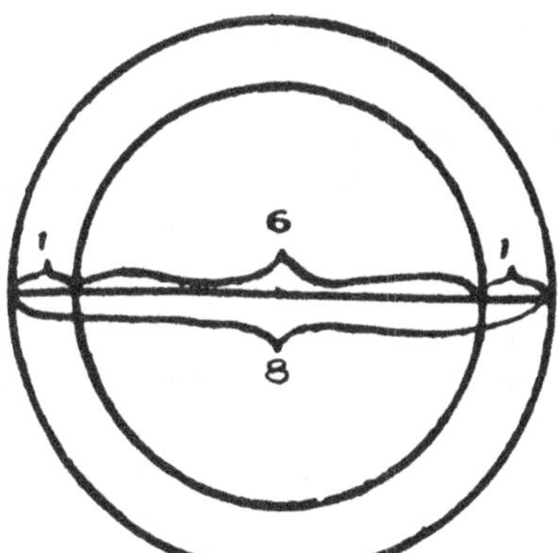

FIG. 6.

These 24 people would then form an annulus (Fig. 6) whose outer circumference is 24 (and therefore whose large diameter is 8) and whose thickness is one (i.e., the space occupied by one person). The diameter of the internal circumference of this annulus would then be 8 less $2=6$ and the internal circumference would consequentlly be 6 times $3 = 18$ cubits. As 18 cubits is more than the minimum $16\frac{4}{5}$ required, therefore R. Yochanan did not mind being somewhat inexact.

NOTE. The whole difficulty in this case would vanish if we assume that R. Yochanan's statement applies to areas instead of linear dimensions. The square described inside a circle whose *area* is 24 square units would (in accordance with Theorem 4, p. 24) be equal to $\frac{2}{3}$ of $24=16$ square units, so that its side is 4 units. R. Yochanan's statement, therefore, is in perfect accord with Rab's statement, i.e., if a booth 4 by 4 square cubits can be inscribed in a circular booth, then the area of the latter $=24$ square cubits.

2. *Size of circular window in a partition wall between 2 courts.*

(חלון עיגול שבין שתי חצרות)

A similar discussion occurs in another place,[1] where the question of the size of a window in a party wall is considered which would legally connect two neighbouring courts for the purpose of allowing the owners to carry things from one area into the other on the Sabbath day. The Mishnah announces that a window 4 by 4 handbreadths fulfils that object provided its lower border is within 10 handbreadths from the ground, and R. Yochanan says in the Gemara that if the window is circular its circumference must be 24 handbreadths, *two and a fraction of which are within 10 handbreadths from the ground.* It is not necessary to enter into the full discussion here, since practically the same arguments are used as in the case of the booth, but here also the difficulties vanish on the assumption that R. Yochanan's statement applies to the area of the circle rather than the length of its circumference. The difference between the area of the circle and that of its inscribed square $=24-16=8$. Therefore the segment of the circle cut off by each side of the square (e.g., the small portion of the circle cut off by the side FG in Fig. 2) $=\frac{1}{4}$ of $8=2$ square hand-

[1] Erubin 76a.

breadths. Hence if 2 and a fraction square handbreadths of the circle are within 10 handbreadths from the ground the lower border of the inscribed square will be within 10 handbreadths from the bottom and will thus fulfil the Mishnah's requirements.

3. The Mensuration of Family Graves. (כוכין)

R. Simeon lays down in the Mishnah[1] the following elaborate plan for constructing family graves (under suitable conditions of the soil).

(a) A Court ABCD (Fig. 7), 6 cubits long, 6 cubits wide (and 4 cubits deep) to accommodate the coffin and its bearers.

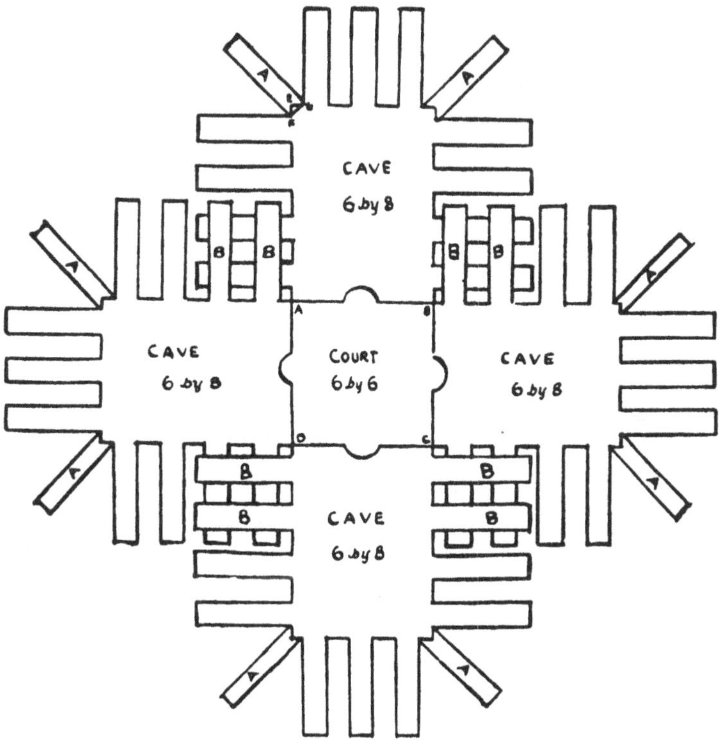

FIG. 7.

(b) 4 Caves opening out of each wall of the court, each cave being 8 cubits long, 6 cubits wide (and 4 cubits deep), and having cubicles for 13 coffins, each cubicle having the following dimensions: 4 cubits long, 1 cubit (or 6 handbreadths) wide, and $1\frac{1}{8}$ cubits (or 7 handbreadths) deep.

[1] Baba Bathra vi, 8.

These cubicles are to be arranged as follows (see Fig. 7):-

4, opening along each long wall of the cave;
3, opening along each distal short wall (i.e., the short wall opposite the one communicating with the Court;
2, opening into each distal corner (marked A).

The first difficulty raised by an anonymous Rabbi in the Gemara[1] is regarding the corner cubicles. On the assumption that the other cubicles (i.e., not the corner ones) open out perpendicularly from the walls (as in Fig. 7) and allowing for a space of one cubit between one cubicle and another, it necessarily follows that the distance between each corner of the cave, and each of its adjacent cubicles along each wall must be $\frac{1}{2}$ a cubit. Thus in the case of the long wall EA, if we allow $\frac{1}{2}$ cubit for the distance EF, between the corner E (for instance) and the first cubicle, then as each of the 4 cubicles occupies one cubit, all 4 cubicles occupy 4 cubits; the 3 intervening spaces between the cubicles, occupy 3 cubits; and the distance between the last cubicle and the other corner being another $\frac{1}{2}$ cubit, we get a total of $\frac{1}{2}+4+3+\frac{1}{2}=8$ cubits = length of wall. Such being the case, asks the Rabbi, "how can we accommodate a cubicle 1 cubit wide in the corner, since the diameter of a square is equal to $1\frac{2}{3}$ of its side?" What he obviously means is that since EF and EG are each $\frac{1}{2}$ a cubit long, therefore GF is the hypotenuse of a right angled isosceles triangle, of sides $\frac{1}{2}$ cubit (i.e., it is the diameter of a square of side $\frac{1}{2}$ cubit) and its length must be $1\frac{2}{5}$ of $\frac{1}{2}=\frac{7}{10}$ cubits. In other words it is less than one cubit and therefore cannot accommodate a cubicle A, which is 1 cubit wide.

The answer given by R. Ashi is that the corner cubicles are placed at a lower level than the others, and therefore GF can be made of any size. Indeed, says R. Ashi, we *must* assume that some cubicles are placed deeper than others, otherwise the first 2 cubicles along each of the neighbouring walls (at right angles to each other) of 2 adjacent caves would intersect (see B in Fig. 7).

R. Huna, the son of R. Joshua, does not accept this solution of differences in level, but offers an alternative solution that the cubicles, instead of opening perpendicularly to the walls, are arranged at angles of 45^0 (חרותא) to their respective walls (see

[1] Baba Bathra 101b.

Fig. 8). In this way the 8 cubicles along the 2 walls, say HB, and BK, at right angles to the 2 neighbouring caves, are parallel and do not therefore intersect even when placed at the same level.

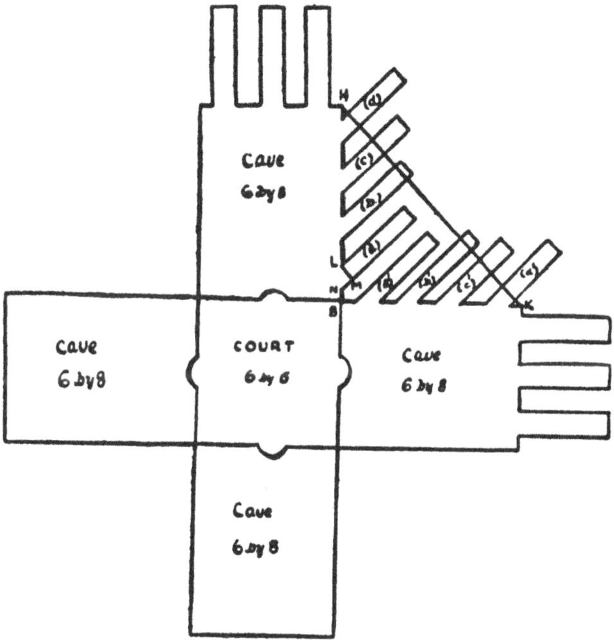

Fig 8

Fig. 8

This solution, however, is rightly rejected by another rabbi. For, as the 2 walls (BH and BK for example) of neighbouring caves form the sides of a square, 8 cubits by 8 cubits, then the diagonal HK is $1\frac{2}{5}$ of $8=11\frac{1}{5}$ cubits, and hence it is impossible to accommodate 8 cubicles each of which is one cubit wide with their 6 interspaces of one cubit each and 4 corner spaces of $\frac{1}{2}$ cubit each, viz., a total of 16 cubits along a diagonal of $11\frac{2}{5}$ cubits. Indeed, as the length LN (for instance) forming the entrance to a cubicle is 1 cubit, then $LM=\frac{1\cdot4}{2}$ so that the width of the cubicle cannot be more than 0·7 cubit.

Another alternative solution is therefore offered, viz., that the cubicles are meant to accommodate children's coffins, which are smaller and therefore would not intersect even if all are placed at the same level and at right angles to the wall. It would also solve the difficulty about the angular cubicle. The size of a child's cubicle is not mentioned.

NOTE Rashi did not complete his commentary to Baba Bathra, and from p. 29a, the task was undertaken by his grandson RABBI SAMUEL BEN MEIR (*Rashbam*). It is therefore of interest to observe that like his illustrious grand. father (See end of this Chapter) he shows a lack of knowledge of elementary geometry. For not only does his diagram (See Fig. 8.) show that all the cubicles cross the diagonol HK, but he goes out of his way to say that they do. (כולם יוצאים באלכסון ויוצאים דרך שם) This of course is not true, since the first cubicles (a) and (a,) will just fail to meet the diagonal. Thus in Fig. 9, RSUL represents the first cubicle along the wall BH. If PQ is perpendicular to the semidiagonal BN then since PQ = $\frac{1}{2}$ cubit, therefore BQ = $\frac{1}{2}$ cubit. Similarly RS = 1 cubit and therefore PS = QT=1 cubit. Therefore BT = $1\frac{1}{2}$ cubits. Therefore BM = 5.5 cubits (Since TM = SL 4 cubits), But BN = $1\frac{1}{2}\cdot 2$ = 5.6 cubits. Therefore, the first cubicle fails (by $\frac{1}{10}$ cubit) to reach the diagonal HK.

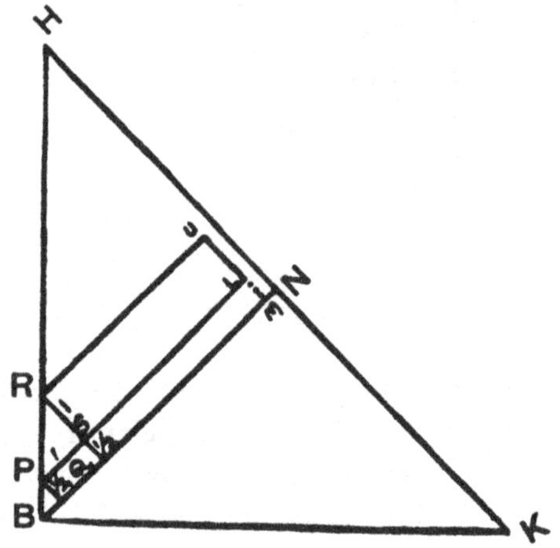

Fig. 9.

The Rashbam also makes the diagonal of a rectangle 4×6 (i.e. $\sqrt{4^2+6^2}$ or $\sqrt{52}$) equal to the diagonal of a square 5 × 5 (i.e. $\sqrt{50}$) and gives a fallacious demonstration in proof of his statement (viz. because the periphery of each of these figures is the same—viz. 20).[1]

[1] Baba Bathra 102a.

4. The Mathematics of Sabbath Distances. (כיצד מעברין את הערים)

(*a*) *Ordinary Towns.* The distance 2,000 cubits which a person may walk in a public thoroughfare from any given private place (רשות היחיד) on the Sabbath day is called the Sabbath distance or "techum" (תחום), and the treatise Erubin deals more especially with the question of methods by means of which one can ritually extend this Sabbath distance—by joining two separate places by some imaginary combination or erub (ערוב) — so as to convert a public thoroughfare into a private court.

Acccording to R. Chanina ben Antigonos[1] the Sabbath distance extends to one of 2,000 cubits radially in all directions (אלפים אמה לכל רוח עגולות) But the sages extend this distance by saying that you circumscribe a square round the circle of 2,000 cubits (Fig. 10) and the permissible Sabbath distance extends to the corners of the square, that is to say that, except for the points of contact between the circle and the square, the *techum* is considerably increased in each direction. Indeed, in the direction of the diagonal it is increased by 40 per cent, taking the ratio between the diagonal and the side of a square as being $1\frac{2}{5}$.

The question is next discussed regarding the Sabbath distance from a town, and the law is laid down as follows:[2] "When the town is circular, you add corners; when square, you do not add corners; if long on one side and short on the other, you consider it square." This means that in the case of a circular town you do not measure the Sabbath distance from the circumference, but from the periphery of its circumscribed square, and the distance is, therefore, extended in the same way as before, and further, "an irregular town is to be considered as a square, of side equal to the longest side." (See Rashi) Further down[3] the distance is still more extended as follows:*

Circumscribe a square ABCD (Fig. 10) round the town, construct squares on each of the sides of this square as ABKL, BCQP, etc., and the Sabbath distances extend to the periphery of the big square HEFG, so that the Sabbath distance (say the Rabbis) is increased by 1,200 cubits each way. This gain, as

[1] Erubin 49b (Mishnah) [2] Erubin 55a. [3] Ib. 56b.

*The sense only of the discussion and not the literal translation is here given.

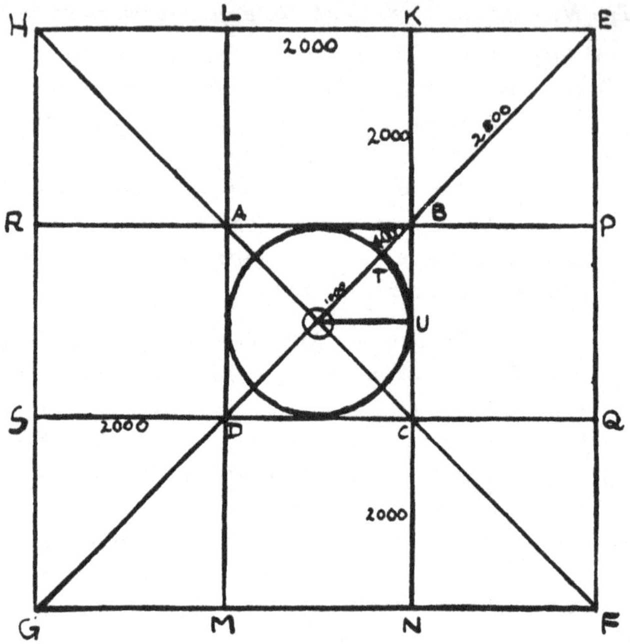

FIG. 10.

Abaye points out is made only when the town itself has a diameter of 2,000 (or a radius of 1,000).

No explanation is given in the Talmud as to how these figures are obtained but the explanation as well as the point in Abaye's observation will be clear from the diagram (Fig. 10).

The Diameter, (which $=$ DC) $= 2{,}000$.
Therefore DB $= 2{,}000 \times 1\frac{2}{5} = 2{,}800$.
Therefore OB $= 1{,}400$.
But OT $= 1{,}000$. Therefore TB $= 400$.
Similarly BE $= 2{,}800$. Therefore TE $= 3{,}200$

Hence the Sabbath distance is extended in the direction OE from 2,000 to 3,200 cubits, i.e., by 1,200 cubits. It is similarly extended in the opposite direction OG, so that altogether it is extended by 2,400 cubits along each of the diagonals GE and HF. It is also obvious that if the diameter of the town is other than 2,000 TB would not be 400, and therefore the total gain would be either more or less than (but not equal to) the figure given.

(β) *Levitical Towns.* (תחום ערי לויים)

The discussion following immediately upon the above relates to certain mensuration problems as well as to the Sabbath distance in connection with the Levitical towns as described in Numbers xxxv, 4, according to which every town (without specification of the dimensions) shall have suburbs which " shall reach from the city and outwards a thousand cubits round about."

[The discussion has here been divided into paragraphs (a), (b), (c), (d), (e), and (f), for easier reference, although there is no such division in the text. The words in inverted commas are as nearly as possible literal translations of the words given in the Talmud, and the words in brackets have been put in by the present author to elucidate the meaning of the phrases preceding them.]

(a) R. Eliezer ben R. Josi says: " The *techum* of the Levitical towns is 2,000 cubits of which 1,000 cubits are suburb or *Migrash* (מגרש): the Migrash is one quarter and the remainder consists of fields and vineyards."

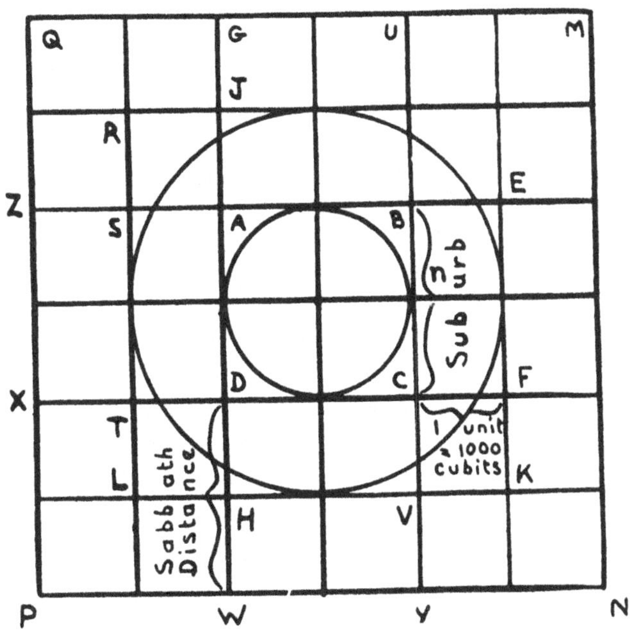

Fig 11.

(b) But surely, asks an anonymous Rabbi, the Migrash is " ½, and not ¼? "

It will be noticed that it is not specified what it is that the Migrash is ½ or ¼ of, but the meaning of this as well as of the subsequent discussion regarding R. Eliezer ben Josi's statement will become clear from the diagram (Fig. 11). The square town ABCD has sides **n** units long (a unit being 1,000 cubits). The subsurbs, therefore (e.g., BCEF, ADST, etc.), each of which is $n \times 1$ square units, have a total area of *4n square units*. The "techum" on each side being according to the questioner the big rectangles (like AGUB), each of which is $n \times 2$ square units, has a total area of 8n square units. Therefore the questioner's contention is correct, since on this interpretation

$$\frac{\text{Migrash}}{\text{Techum}} = \frac{4n}{8n} = \frac{1}{2}$$

(c) Then Raba bar Ada replies that "the discussion relates to a town 2,000 by 2,000" (i.e., 2 by 2 units). "But if this is the case," proceeds the questioner, "then *the whole techum* (i.e., the 4 rectangles like AGUB) has an area of 16 (since $8n = 16$), the corner squares (each of which like AGQZ being 2 by 2) have a total area of 16, (so that the techum and corner squares together, i.e., $8n + 16$, have an area of *32 square units*). "Take away 8" (i.e., 4n of the Migrash) "from the techum and also subtract 4 from the corners", (viz., the 4 small squares like AJRS, each of which is 1 by 1) i.e., take away altogether $4n + 4$ or 12 square units, and we then get the

$$\frac{\text{Migrash}}{\text{Techum}} = \frac{4n+4}{8n+16} = \frac{12}{32} = \frac{3}{8}$$

"and this" points out the questioner "is greater than ⅓" (instead of being ¼ as originally stated).

(d) Then somebody answers that the area of the town itself (i.e., n^2, which $= 4$ square units) must also be taken into consideration (so that the denominator is not $8n + 16$, but $n^2 + 8n + 16$, i.e., the whole square QMNP).

(e) "But," asks the questioner again, "in that case the proportion surely is ⅓?" i.e.,

$$\frac{4n+4}{n^2+8n+16} = \frac{8+4}{4+16+16} = \frac{12}{36} = \frac{1}{3}$$

[It will be noted that on this interpretation of numerator and denominator, it is possible to find the dimensions of the town, such that the ratio $=\frac{1}{4}$, as given originally by R. Eliezer ben Josi. Thus

$$\text{if} \quad \frac{4n+4}{n^2+8n+16}=\frac{1}{4}$$

$$\text{Then } 16n+16=n^2+8n+16$$

$$\text{Therefore } n^2-8n=0, \text{ whence } n=8.$$

So that for a town 8,000 by 8,000 cubits, the proportion as given originally would hold good, thus

$$\frac{4n+4}{n^2+8n+16}=\frac{32+4}{64+64+16}=\frac{36}{144}=\frac{1}{4}$$

Indeed Tosefoth raises this question, but answers that the Levical cities were not so large.]

(f) The answer to (e) as given in the Talmud is that " the shape of the town here considered is not square but circular (of diameter 2,000 cubits). In that case the Migrash consists of the annulus between the 2 circles, and as these circles have radii of 2 and 1 units respectively, therefore the area of the annulus is $\pi(2^2)-\pi(1)^2=9$ square units (since $\pi=3$). And as the denominator is the whole square QMNP $=36$, therefore Ratio

$$\frac{\text{Migrash}}{\text{Whole area}}=\frac{9}{36}=\frac{1}{4}$$

The Talmud continues the discussion in which different people give different interpretations to the numerator and denominator. Thus Abaye takes the numerator (Migrash) as the 4 rectangles of dimensions n by 1 together with their adjacent squares of 1 by 1, and for denominator (techum) he takes the large rectangles of n by 2 and their adjacent squares of 2 by 2. He further supposes the town to be circular, of diameter 1,000 cubits. The ratio then becomes

$\frac{\text{Migrash}}{\text{Techum}}=\frac{3}{4}\left(\frac{4n+4}{8n+16}\right)$ since the area of a circle is $\frac{3}{4}$ of that of the circumscribed square

$$=\frac{3}{4}\left(\frac{4+4}{8n+16}\right)=\frac{3}{4}\times\frac{1}{3}=\frac{1}{4}$$

Rabina takes the town to be a square 2,000 by 2,000; the Migrash he calls the small rectangles on its sides $=4n$, and as the denominator he takes $8n+16$, so that $\frac{4n}{8n+16}=\frac{8}{32}=\frac{1}{4}$

R. Chanilai of Chusnah, who was obviously no mathematician, raises certain difficulties with regard to the extension of the Sabbath distance into which it is not necessary to enter here, but in which he shows that he erroneously applied the proportion $\frac{\text{area of circle}}{\text{area of circumscribed square}}$ (which the Talmud so frequently gives as $\frac{3}{4}$), to the proportion $\frac{\text{side of square}}{\text{diagonal}}$ In other words he makes the diagonal $= 1\frac{1}{3}$ of the side (cf. p. 54). In this he is corrected by R. Ashi, who points out that the diagonal is $1\frac{2}{5}$ of the side.

5. *Measurement of Depths, Heights, and Distances.*

(a) *Depths.*

I have already mentioned (p. 11) that an instrument resembling an astrolabe by means of which one could measure a distance of 2,000 cubits, is mentioned in the Talmud.[1] The following is a description of the method of using it. "To find the depth of a valley, one should bring a tube and look through it on dry land and he will find the depth of the valley." This is very obscure, but Rashi explains it as follows: "The observer should first focus or adjust the apparatus. Let him then go to the edge of the valley and look down into its deepest part. Let him walk back until he comes to a place from which he can just see the bottom of the valley (that is to say, if he were to walk a little further back he would no more see it). He will then know that the depth of the valley and his own distance from the edge of the valley are as the distance of the vision of the tube."

Maimonides *in loco* says that this passage can only be understood by those who know all about trigonometry and solution of triangles.[2] Although Rashi's meaning is somewhat obscure the calculation can be made from his directions as follows (see Fig. 12):-

Let h be the height of the man AB or CD (up to his eyes); H, the depth of the valley FG; AC the known distance of the 2nd position of the man from the edge of the valley. Then

[1] Erubin 43b.

[2] לא יבין זה אלא מי שידע בחכמת התשבורת, וידע מה שיתחייב מן היחס בדמיונות חרוטין וזה לזה ואיך יצא הצלע שאינו ידוע, כשתדע צלעות אחרות. פירוש המשניות להרמב"ם (מ"א:).

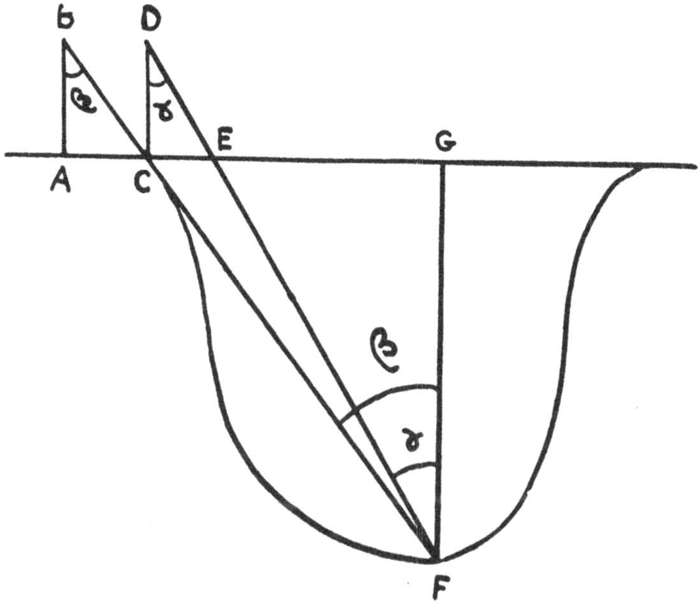

FIG. 12.

$$CE = CD \tan \alpha = h \tan \alpha$$
$$CG = FG \tan \beta \quad = H \tan \beta$$
$$\text{and } EG = CG - CE = H \tan \beta - h \tan \alpha$$
$$\text{But } EG = H \tan \alpha$$
$$\text{Therefore } H \tan \alpha = H \tan \beta - h \tan \alpha$$
$$\text{Therefore } H (\tan \beta - \tan \alpha) = h \tan \alpha$$
$$\text{Therefore } H = \frac{h \tan \alpha}{\tan \beta - \tan \alpha}$$

(β) *Heights.*

" To find the height of a tree, it is necessary to measure one's own height, the length of one's shadow, and the length of the tree's shadow."[1]

From Fig. 13 it is obvious that if AB is the tree, AE its shadow, CD the man, and CE his shadow, then

$$\frac{CD}{AB} = \frac{CE}{AE} \quad \therefore AB = \frac{CD.AE}{CE}$$

[1] Erubin 48b.

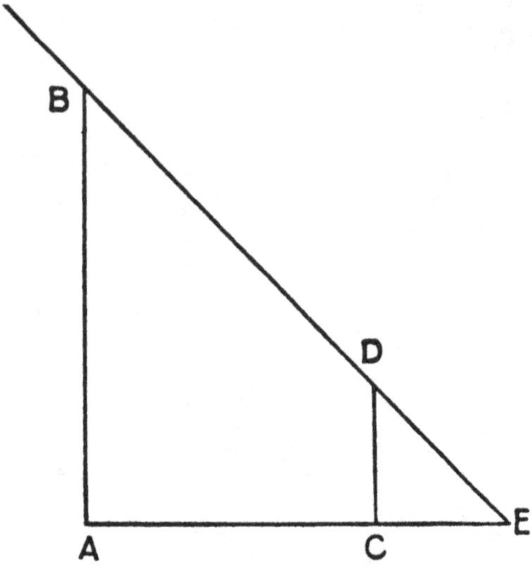

Fig. 13.

Plutarch states that the Greek Thales (640-546 B.C.) of Miletus, who introduced the study of Geometry into Greece, amazed King Amasis by measuring the heights of pyramids from their shadows. [1]

(γ) *Estimation of Distances.*

The Mishnah[2] narrates that on a certain occasion R. Gamliel together with three other Rabbis arrived at a certain place on Friday evening and he decided that it was permissible to land as " observations showed that the harbour was within the Sabbath distance (תחום) of 2,000 cubits." The method used by R. Gamliel is indicated in the Palestinian Talmud where it is stated that he knew the heights of certain towers along the coasts to the tops of of which he used to direct his telescope for the purpose of estimating the distance at which he was removed from them. [3]

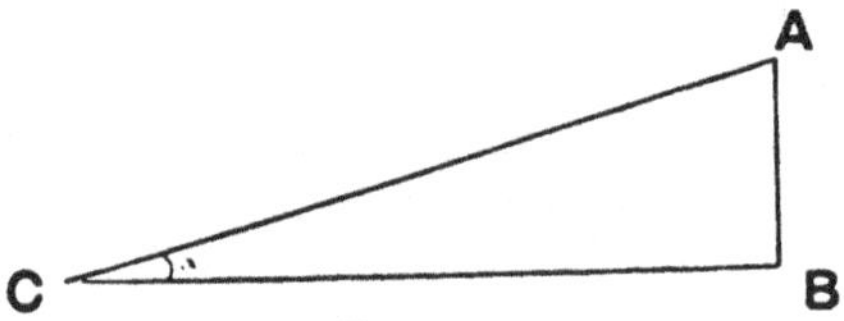

Fig. 14.

Suppose the observer to be at C (Fig. 14), and the height of AB is known as well as the angle ACB made by the telescope when pointing to A.

[1] See Cajori, op. cit. [2] Erubin iv. 2.

[3] מצודות היו לו לרבן גמליאל, שהיה משער בהן עיניו במישר.

$$\text{Then } \frac{AB}{CB} = \tan ACB \text{ or } CB = \frac{AB}{\tan ACB}$$

Thus if AB = 150 cubits, and angle ACB = 4°. 17′

$$CB = \frac{150}{\tan 4^\circ\ 17'} = \frac{150}{0{\cdot}0752} = 1995{\cdot}7 \text{ cubits}$$

so that the distance of the coast is less than the permissible 2,000 cubits.

(8) *Estimation of the length of the base of a hill.*

The Mishnah[1] gives details of the method of measuring the 2,000 cubits within which a person is allowed to walk on the Sabbath day (תחום) If a hill is encountered which is too high or whose base is too long for the measuring tape, (which must be exactly 50 cubits long), to be passed over the summit of the hill, by means of long poles planted in the ground at the foot of the hill on the two opposite sides, then, says the Mishnah, the hill must be "tunnelled through." היה מודד והגיע להר,.. אם אינו יכול להבליעו... מקדרין.

The gemara[2] explains that the term "tunnelled through" is to be understood in a figurative sense, and means that in such a case a special form of measurement must be undertaken as follows: One surveyor A of average height, stands at the foot of the hill and holds one end of a tape measure, 4 cubits long, against his heart (or pit of the stomach) and a second surveyor B, holding the other end of the measure, walks up the hill until the tape stretches tightly between his own feet and A's heart. A next walks up to B, and the latter proceeds higher up on the hill until the tape measure again stretches tightly between A's heart and B's feet. The process is repeated until the whole of the hill is covered. Then by knowing the slope of the hill, i.e., the angle which the surface of the hill makes with the horizontal, the length of its base can be calculated.

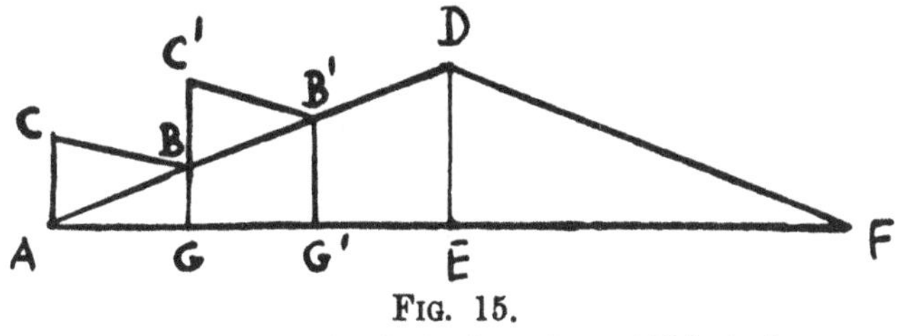

FIG. 15.

Let ADF (Fig. 15) be the hill whose base AF is to be measured. The first surveyor stands at A holding the end of the 4 cubits

[1] Erubin v. 4. [2] Ibid 58b.

measure near his heart at C. The height of the heart's apex in a man of average height is known to be 2 cubits.[1] Therefore AC = 2 cubits. Also CB, the length of the tape between A's heart and B's feet = 4 cubits, and the angle CAB = 90 − angle DAE, is also known, since the angle DAE has been previously determined, presumably by the optical instrument שפופרת already referred to. We, therefore, have a triangle ACB of which the 2 sides AC, CB, and their contained angle are known. Therefore the triangle is completely solved and AB is determined. Hence in triangle ABG, we have: AB is known and the angle BGA = 90° and angle BAG is known. Therefore the side AG is determined. Similarly the length of BB′ is determined and thence the length of BH or of GG′ is calculated. In this way the whole length of AF can be found.

The Talmud says that this method is to be adopted only if Sin DAE is greater than $\frac{1}{3}$ — say $\frac{5}{12}$

(2) אמר רבא לא שנו אלא בהר המתלקט י' (טפחים) מתוך ד' (אמות) אבל
בהר המתלקט י' (טפחים) מתוך ה' (אמות) מודדו מדידה יפה.[2]

Suppose Sin DAE $= \frac{5}{12} = \cdot 4167$, ∴ DAE $= 24^\circ\ 38'$ (to the nearest minute).

$$\therefore \text{CAB} = 90 - 24^\circ\ 38' = 65^\circ\ 22'.$$

But from Trigonometry it is known that

$$\frac{AC}{CB} = \frac{\text{Sin CBA}}{\text{Sin CAB}}, \text{ i.e. } \frac{2}{4} \text{ or } \frac{1}{2} = \frac{\text{Sin CBA}}{\text{Sin } 65^\circ.22'}$$

$$\therefore \text{Sin CBA} = \tfrac{1}{2} \text{ Sin } 65^\circ.22' = \tfrac{1}{2} \times 0\cdot9090 = 0\cdot4545.$$

$$\therefore \text{CBA} = 27^\circ.\ 2'. \quad \therefore \text{ACB} = 180^\circ - (65^\circ.\ 22' + 27^\circ.\ 2') = 87^\circ.\ 36'$$

$$\text{But } \frac{AB}{AC} = \frac{\text{Sin ACB}}{\text{Sin ABC}} = \frac{\text{Sin } 87^\circ.\ 36'}{\text{Sin } 27^\circ.\ 2'} = \frac{9991}{4545} = 2\cdot1983$$

$$\therefore AB = AC \times 2\cdot1983 = 4\cdot3966$$

$$\text{But AG} = \text{AB Cos BAG} = 4\cdot3966 \text{ Cos } 24^\circ.\ 38'$$
$$= 4\cdot3966 \times \cdot9090 = 3\cdot9965 \text{ cubits.}$$

Hence every 4 cubits along CB corresponds to
3·9965 cubits along the base.

The reason why this method of measurement is not to be adopted if Sin DAE is less than $\frac{1}{3}$ is that in such a case the hill is low

[1] Mechilta to Exod. xvi. 13.

[2] A cubit contains 6 handsbreaths, therefore a ratio of $\frac{10 \text{ handsbreadths}}{4 \text{ cubits}} = \frac{10}{24} = \frac{5}{12}$ and one of $\frac{10 \text{ handsbreadths}}{5 \text{ cubits}} = \frac{10}{30} = \frac{1}{3}$

enough for the ordinary measuring tape of 50 cubits to be stretched over its peak by means of two poles of reasonable length planted at the edges of the hill on the opposite sides—provided the base of the hill is not longer than 50 cubits.

Thus, in the figure, if $AF=50$, then $AE=25$, and as $ED=\frac{1}{3}AD$, or $AD=3ED$, we get

$$AD^2=ED^2+AE^2$$
$$\text{i.e., } 9ED^2=ED^2+AE^2$$
$$\therefore\ AE^2=8ED^2$$
$$\therefore\ ED=\frac{AE}{\sqrt{8}}=\frac{25}{2\sqrt{2}}=\frac{25}{2\times 1\cdot 4}=8\cdot 9 \text{ cubits}$$

and poles about 9 cubits high are not too long to be used If either AE is less than 25, or Sin DAE is less than $\frac{1}{3}$, then ED is less than 8·9 cubits; e.g., if $AE=20$, then $ED=\frac{20}{2\cdot 8}=7\cdot 14$ cubits. Similarly, if Sin $DAE=\frac{1}{4}$, then $ED=\frac{AE}{\sqrt{15}}=\frac{25}{3\cdot 87}=6\cdot 5$ cubits.

If, however, $\frac{ED}{AD}=\frac{5}{12}$ then $AD=\frac{12ED}{5}$

$$\therefore\ AD^2 \text{ or } \frac{144}{25}ED^2=ED^2+EA^2$$
$$\therefore\ AE^2=\frac{119}{25}ED^2,\text{ whence } ED=\frac{5AE}{\sqrt{119}}=\frac{125}{\sqrt{119}}$$

i.e., ED is greater than 11 cubits, and poles so high cannot be conveniently used.

6. *Agricultural Mathematics.* (כלאים)

(a) The Mishnah[1] gives the following rules (for evading the Mosaic prohibition (Levit. xix, 19) of sowing a field with mingled seeds). A square furrow 6 × 6 handbreadths may be sown with 5 kinds of seeds, viz., one on each of the sides and one in the middle. If the furrow is fenced round with a fence one handbreadth high and one handbreadth deep, then there may be 13 kinds of seeds sown, viz., 3 on each of the sides and one in the middle. R. Judah allows 6 in the middle (Cf. Sabbath 84b, Pesachim 39b).

The Mishnah does not state the exact manner of arranging the furrows, but it is assumed that the furrows must not touch one

[1] Kilayim iii, 1.

another at more than one point, in order not to constitute "mingling" in the Biblical sense.

The arrangements as shown in the diagrams (Figs. 16 to 19), of the sown areas (which are not described in the Mishnah or Gemara), are based on those suggested by Maimonides, although he does not state the manner in which he made his calculations.

1st case (Fig. 16) Let $PQ = RS = UT = VW = x$ handbreadths and $AP = QB = AR$ etc., $= y$ handbreadths.

Hence area of each sown furrow $= xy$,

and all sown furrows $= 4xy$

The middle sown area $= \frac{x^2}{2}$

Therefore whole sown area $= 4xy + \frac{x^2}{2}$

But $AB = 6$. Therefore $x + 2y = 6$.

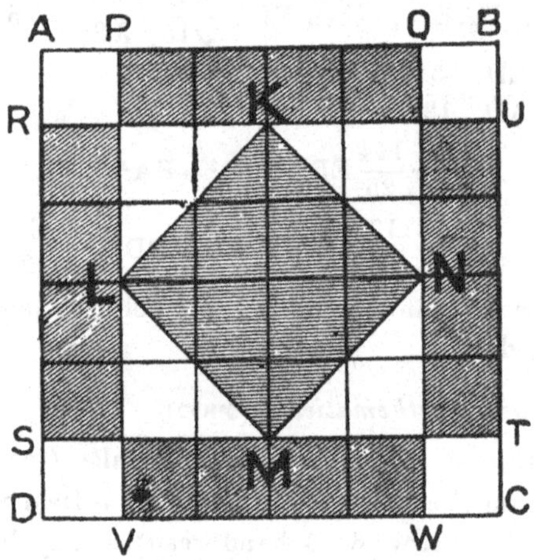

FIG. 16.

Therefore $y = \frac{6-x}{2}$

Therefore whole sown area $= \frac{4x\ (6-x)}{2} + \frac{x^2}{2}$

Let this $= \mathbf{a}$

Therefore $\frac{4x(6-x)}{2} + \frac{x^2}{2} = a$

i.e. $24x - 4x^2 + x^2 = 2a$

Therefore $3x^2 - 24x + 2a = 0$

The roots of this equation are $x = 4 \pm \sqrt{16 - \frac{2a}{3}}$ As a is positive, therefore the greatest value it can have is such as to make $16 - \frac{2a}{3} = 0$ i e. $a = 24$, Hence the largest sown area is 24, and with this value of a, $x = 4$ and $y = \frac{6-x}{2} = 1$.

Hence, whole area of piece of ground i.e., $(x+2y)^2 = 36$, and sown area (i.e., $4xy + \frac{x^2}{2}$) $= 24$, i.e., $\frac{2}{3}$ of the whole plot. This is the value given by Maimonides without (as I have mentioned) stating the manner in which it was calculated.

See also the special commentary of R. Obadya of Bertinoro (died 1510) on the same Mishnah.

2nd case (b) Let ABCD (Fig. 17) $= 6 \times 6$, EFGH being the surrounding raised fence. On this fence are twelve different furrows of 1 square handbreadth each, viz., three on each side at fixed

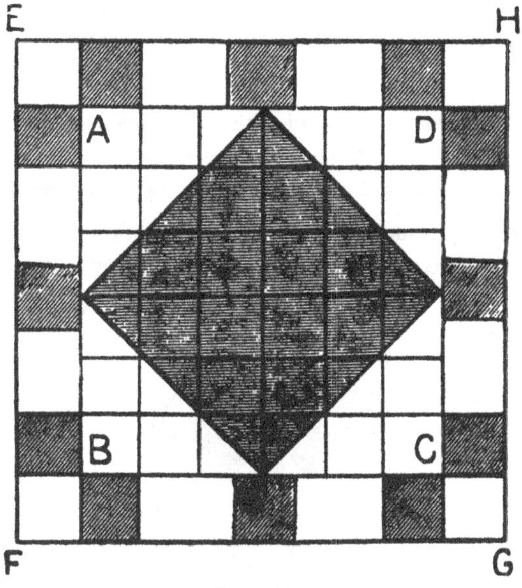

FIG. 17.

distances from one another, and there is one on the plot itself, the latter covering 18 square handbreadths.

Therefore total area of sown parts $= 30$ square handbreadths (see Fig. 16) $= \frac{15}{32}$ of whole (since whole $= 64$ sq. handbreadths).

FIG. 18.

In order to satisfy R. Judah's statement, imagine Fig. 18 where each of the six kinds of seeds in the middle of the plot covers a rhombus whose area obviously $=3$ square handbreadths (since it is inscribed in a rectangle of area 2×3).

Therefore the whole covered area $=30$ square handbreadths. The problem then arises, Is the distance between each two parallel rhombi the *required* minimum distance of $1\frac{1}{2}$ handbreadths?

The Rambam (Maimonides), who gives the above construction, erroneously takes GD (Fig. 19) as the required distance which

$$=\sqrt{DK^2+KG^2}=\sqrt{(1\tfrac{1}{2})^2+1^2}=\sqrt{\tfrac{9}{4}+1}=\tfrac{1}{2}\sqrt{13}=1\tfrac{4}{5}$$

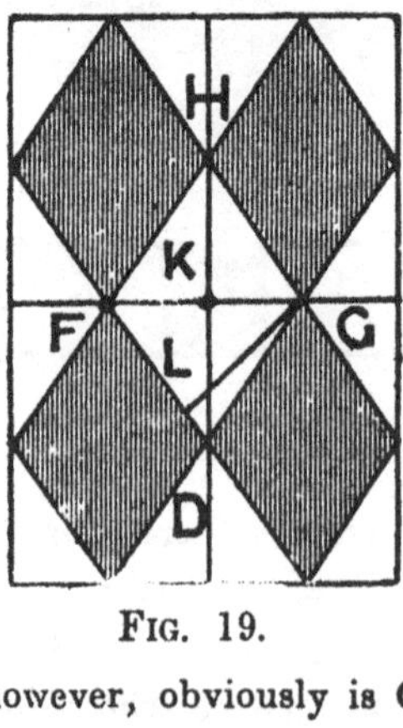

FIG. 19.

The real distance, however, obviously is GL in Fig. 19 (where

GL is perpendicular to FD), and GL = GD sin GDF.

It is easy to find the value of the angle GDF,

For $\angle GDF = 2\angle KDG$.

But $\tan KDG = \frac{1}{1\frac{1}{2}} = \frac{2}{3} = \cdot 6667$

Therefore $KDG = 33°.\ 41'.\ 30''$

Therefore $GDF = 2KDG = 67°\ 23'$.

Therefore GL, which = GD sin GDF,

$= \frac{1}{2}\sqrt{13}\ \sin\ 67°\ 23' = 1.662$

This is still greater than the minimum $1\frac{1}{2}$, and therefore, although Maimonides was slightly wrong in his calculation, yet the arrangement of the plants as given in his figure is quite correct.

(β) Basing itself on Deut. xxii, 9, the Mishnah[1] makes the statement that, if a green vegetable is planted in a vineyard wherein the vines are 4 or 5 cubits distant from one another, then 45 vines become holy through it as a result of *kilayim* (i.e., mingled seeds).

Maimonides deduces from this the following corollary: If a vegetable is planted in the centre of a vineyard consisting of nine equidistant rows, the distance between the rows being 4 cubits from one another, each row containing nine vines at distance of 4 cubits from one another, then a circular area having the plant as centre and a radius of 16 cubits becomes holy through *kilayim*.

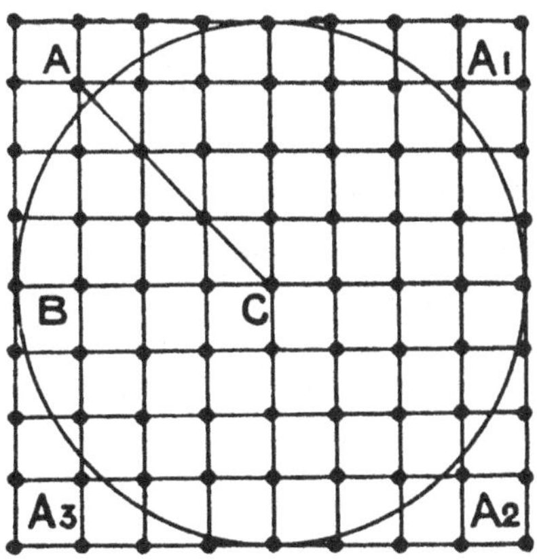

FIG. 20.

[1] Kilayim v. 5.

Tosefoth Yom Tob (Comments of Yom Tob Lipman Heller, of the 17th century) proves this as follows:-

In the diagram (Fig. 20), CB = BA = 12 cubits.
$AC^2 = 2 \times 144 = 288$.
Therefore $AC = \sqrt{288}$, i.e., just less than 17.

Hence the circle with C as centre, and radius = 16 cubits, i.e.; the inscribed circle, will just miss the four vines A, A_1, A_2, A_3, and will also exclude 32 other vines on the tangent of the circle, leaving only 81–36 = 45, i.e., the number specified in the Mishnah.

(7) *Solid Geometry.*

The volume of a prism is given as equal to the product of the area of the base by the height; and that of a cylinder as $\frac{3}{4}$ the product of its diameter by its height (i.e., $\pi r^2 h$ where $\pi = 3$).[1] These statements occur in a discussion on:-

The shape and capacity of the 'Sea of Solomon', which according to 1 Kings vii, 23, 26, had a circumference of 30 cubits, a diameter of 10 cubits, a depth of 5 cubits, and capacity of 2,000 bath'.[2] The discussion opens as follows:-

"R. Chiya taught that the 'Sea of Solomon' had a capacity of 150 Mikvahs."[3] An anonymous questioner then asks: "Since the Sea of Solomon' was a cylinder of 10 cubits diameter and 5 cubits height, its capacity should be $3 \times (\frac{10}{2})^2 \times 5 = 375$ cubic cubits = 125 Mikvahs and not 150 as given by R. Chiya?" To this somebody replies in the name of Rami bar Ezekiel that the 'Sea of Solomon', was not cylindrical throughout, but only for the top 2 cubits, the bottom 3 cubits being prismatic.

Its volume would then be:-

(i) For the prismatic portion, 300 cubic cubits.
(ii) For the cylindrical portion, $3 \times (\frac{10}{2})^2 \times 2 = 150$ cubic cubits.

Total = 450 cubic cubits = 150 Mikvahs or 6,000 Seah or 2,000 baths'.

[1] Erubin 14b.

[2] A 'bath' (בת) is a fluid measure equal to 3 Seah. 2,000 'baths' are therefore equivalent to 6,000 Seah.

[3] A Mikvah (מקוה) is 3 cubic cubits and contains 40 Seah, therefore 150 Mikvahs = 6,000 Seah.

This solution can, of course, be obtained by means of a simple equation, which, it is reasonable to assume, was the method adopted by Rami bar Ezekiel to arrive at his answer.

If x = depth of prismatic portion, then $(5-x)$ = depth of cylindrical portion.

Therefore volume of Sea $= 100x + 75(5-x) = 450$ cubic cubits

Therefore $25x = 75$ cubits, whence $x = 3$ cubits.

NOTE. As the value of π is 3·1416 and not 3, it is obvious that even if we assume the ' Sea of Solomon ' to have the shape just described, it must have contained $300 + (3.1416 \times 50) = 457$ cubic cubits, which is more than 2,000 ' baths ' explicitly given in the Bible, and Zuckermann,[1] in order to reconcile the Biblical statements about the size of its circumference, its diameter and its exact cubical contents, with Rami bar Ezekiel's solution, offers the ingenious explanation that while the upper 2 cubits had a

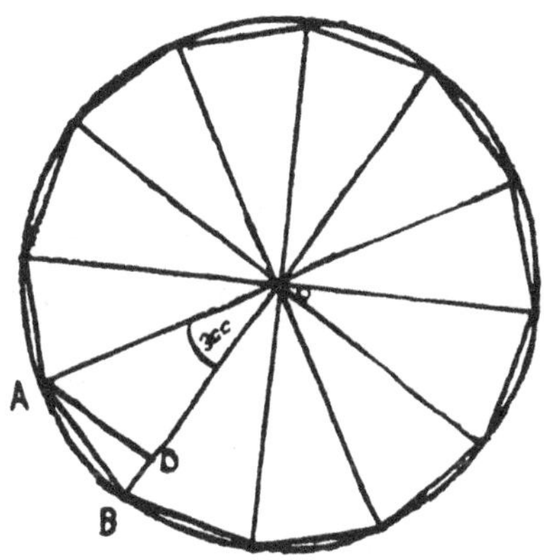

FIG. 21.

cylindrical shape, it was its external surface only that was cylindrical, the internal surface being dodecagonal (see Fig 21). The area of a cross-section of such a vessel would, of course, be equal to 12 times each of its component triangles. Now if in Fig. 21 r = radius of circle and AD is perpendicular to BO, then

$AD = r \sin 30^0 = \frac{1}{2}r$. But area of triangle $AOB = \frac{1}{2}AD.BO = \frac{1}{4}r^2$.

Therefore, area of dodecagon $= 12.\frac{1}{4}r^2 = 3r^2$.

[1] op. cit. p. 29.

Hence in this case (r being 5), the area of the dodecagon $=75$ square cubits and the volume of the upper portion $=75 \times 2=150$ cubic cubits.

(*d*) *Surds.*

The value of $\sqrt{2}$

We have already seen that the diagonal of a square is generally accepted in the Talmud as being $1\frac{2}{5}$ of its side. In other words $\sqrt{2}=1{\cdot}4$.

We have also mentioned that Tosefoth points out that this value is a little less than the real value. In view of this, the following extract from a Mishnah is of interest.

Speaking in connection with Numbers xix, 14, according to which, every one that comes into contact with a tent where there is a dead body or portion of a dead body becomes ritually unclean, and referring to Mishnah III., 7, (Oholoth) where it is stated that a space $1 \times 1 \times 1$ handbreadths constitutes a tent for the purpose, the Mishnah[1] says that "A cylindrical beam lying in an open field and having a circumference of 24 handbreadths fulfils the condition of a tent if a portion of dead body lies immediately under it."

עמוד שהוא מוטל לאויר אם יש בהקיפו כ"ד טפחים מביא את הטומאה תחת דפנו.

If we adopt the relationship between the diameter and the side of a square as given before, viz., Diameter $=1\frac{2}{5}$ side, we must come to the conclusion that the Mishnah's requirement with respect to the circumference of the beam is too large, since with such a circumference, the square space HGBK which would constitute a tent in the Biblical sense would have sides of $1\frac{1}{7}$ by $1\frac{1}{7}$ and not 1 by 1, and indeed the requisite circumference to make HGBK a square of 1 by 1 is not more than 21 cubits. This is seen from the following (see Fig. 22):-

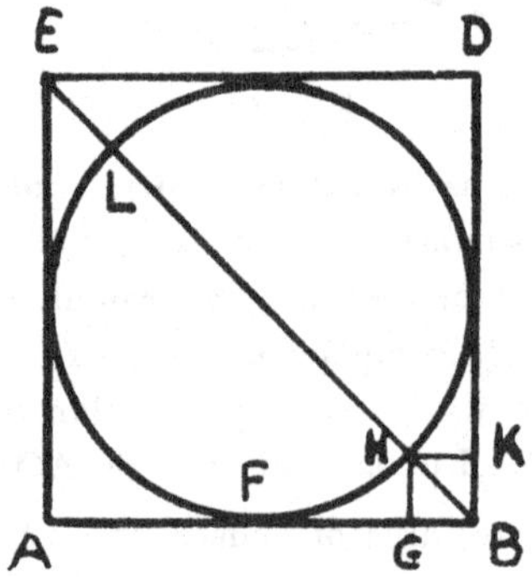

FIG. 22.

[1] Oholoth xii. 7.

If we take the circumference of the beam as 24, then the diameter of the beam LH as well as the side of the circumscribed square (say AB) must be 8 cubits.

We therefore have

(i) $EB = 8 \text{ times } 1\frac{2}{5} = 11\frac{1}{5}$

But $LH = 8$

Therefore $EL + HB = 11\frac{1}{5} - 8 = 3\frac{1}{5}$

i.e., $2HB = 3\frac{1}{5}$, (since $EL = HB$)

Therefore $HB = 1\frac{3}{5}$ or $\frac{8}{5}$

Therefore HG or $GB = \frac{5}{7} \times HB = \frac{5}{7}.\frac{8}{5} = \frac{8}{7} = 1\frac{1}{7}$

(ii) On the other hand if we make HG or $GB = 1$ (in accordance with the Biblical minimum)

Then $HB = 1\frac{2}{5}$

Therefore $EB = LH + 2\frac{4}{5}$

But $EB = 1\frac{2}{5}.EA = 1\frac{2}{5}.LH$

Therefore $LH + 2\frac{4}{5} = 1\frac{2}{5}.LH$

Therefore $2\frac{4}{5} = \frac{2}{5}.LH$

Therefore $LH = 7$

Therefore circumference of beam $= 3LH = 21$

It is, however, possible that the Mishnah in making the circumference of the beam 24 cubits, was not wrong in its calculation but took a smaller value for $\sqrt{2}$. Let us therefore see what value of $\sqrt{2}$ will make a beam 24 cubits in circumference fulfil the Biblical condition with regard to a tent 1 by 1.

In the diagram (Fig. 22)

$EB = AB\sqrt{2} = 8\sqrt{2}$

i.e. $2HB + 8 = 8\sqrt{2}$

But $HB = \sqrt{2}$

Therefore $2\sqrt{2} + 8 = 8\sqrt{2}$

Therefore $\sqrt{2} = \frac{4}{3} = 1.33$

The Mishnah, therefore, may have taken the smaller value of $1\frac{1}{3}$ for $\sqrt{2}$ (instead of that specified in the Talmud, viz., $1\frac{2}{5}$).

Historically it is of interest to note that the Section Taharoth, to which Masechta Oholoth belongs, is one of the older portions of the Mishnah.[1] The later Mishnahs in Erubin were acquainted with the more exact value of 1·41 (see p. 54) although for practical purposes the Rabbis were satisfied with the value of 1·4, or $1\frac{2}{5}$.

[1] Fränkel, Darke Hamischnah (quoted by Zuckermann op. cit. p. 11)

Another point of interest is that this value of $1\frac{1}{3}$ for the relationship between the diagonal and the side of a square is explicitly given by one of the rabbis (R. Chanilai) in the Talmud (see page 40).

A very close approximation to the value of $\sqrt{2}$ can be found indirectly from the statement in the Mishnah that the length of the side of a square whose area is 5,000 square cubits is equal to 70 *and a fraction.* (שיריים)[1] The value of this fraction (Shirayim) is subsequently given in another Mishnah, indirectly, as $\frac{2}{3}$, since it is stated that twice the side of such a square is $141\frac{1}{3}$ cubits long. From this we get therefore that $\sqrt{5000} = 70\frac{2}{3}$, i.e. $50\sqrt{2} = \frac{212}{3}$

$$\text{Whence } \sqrt{2} = \frac{212}{150} = \frac{106}{75}$$
$$= 1{\cdot}4133$$

Note : $1\frac{31}{75}$ is one of the Archimedean approximations of $\sqrt{2}$)

$\sqrt{10}$. Tosefoth to Erubin 5a says " that the diagonal of a rectangle of 3 by 1 $(= \sqrt{3^2 + 1^2} = \sqrt{10})$ is greater than 3 by less than the diagonal of a square of 1 by 1 is greater than 1 ".[2] In other words, $\sqrt{10}$ is less than $3\frac{2}{5}$. The real value is not given, but is, of course, 3·162...

R. Josi says that if the diagonal of a rectangle whose area is 5,000 square cubits) is twice one of the sides, it is permissible to carry within that space.[3] Maimonides makes the approximate dimensions of such a space $93\frac{1}{21}$ cubits and $53\frac{1}{3}$ cubits respectively. Such a space, he says, would have a diagonal of $107\frac{1}{2}$ cubits.

Obviously there is some error in Maimonides' calculation, since not only is $107\frac{1}{2}$ greater, by $\frac{5}{6}$ of a cubit, than twice $53\frac{1}{3}$, but the diagonal of a rectangle $93\frac{1}{21}$ by $53\frac{1}{3} = \sqrt{(93\frac{1}{21})^2 + (53\frac{1}{3})^2} = 107\frac{1}{5}$ and not $107\frac{1}{2}$. The exact dimensions are found as follows :-

Let sides of rectangle be x and y (Fig. 23).

$\therefore \quad xy = 5{,}000$ and $\sqrt{x^2 + y^2} = 2x$ (by hypothesis).

$\therefore \quad x^2 + y^2 = 4x^2$, whence $y = \sqrt{3}$, and $xy = x^2\sqrt{3}$

$\therefore \quad x^2\sqrt{3} = 5{,}000$, whence $x = \dfrac{\sqrt{5{,}000}}{\sqrt[4]{3}} = 53{\cdot}7285$

and $y = x\sqrt{3} = 53{\cdot}7285 \times 1{\cdot}732 = 93{\cdot}0605$

$\therefore \quad \text{diagonal} = \sqrt{53{\cdot}7285^2 + 93{\cdot}0605^2} = 107{\cdot}457.$

Possibly, the $53\frac{1}{3}$ of Maimonides is a misprint for $53\frac{2}{3}$.

[1] Erubin ii. 5.
[2] See Commentary of R. Samuel Edels of the 17th Century (Maharshah) ad loc.
[3] Erubin ii. 5.

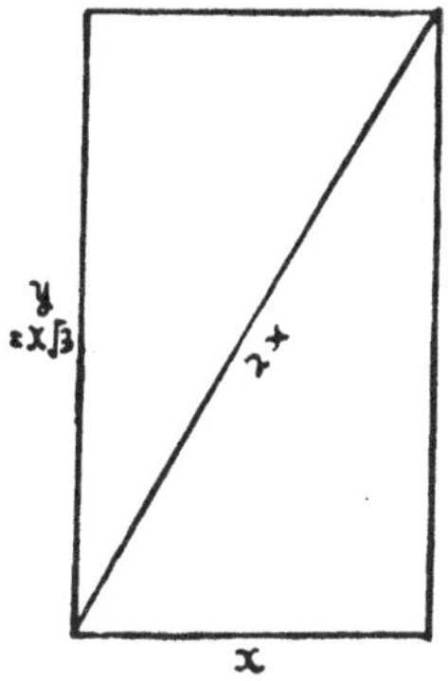

FIG. 23.

Incidentally, we see that Maimonides makes the value of $\sqrt{3}$ $\left(\text{i.e. } \frac{y}{x}\right) = \frac{93\frac{1}{27}}{53\frac{1}{3}} = 1{\cdot}74\ldots$ instead of its true value of $1{\cdot}73\ldots$ If, however, $53\frac{1}{3}$ is a misprint for $53\frac{3}{5}$, then the value of $\sqrt{3}$ comes out as $1{\cdot}73\ldots$ which is correct to 2 decimal places.

$\sqrt{5000}$. The question of $\sqrt{5000}$ arises in connection with the discussion of the dimensions of an enclosed space within which one is allowed to carry things on the Sabbath day. The law is that in a *Beth Sothaim* (בית סאתים) i.e., inside a fenced-in space of any shape, regular or irregular, square, rectangular, circular, triangular, etc., having an area of 5,000 square cubits, carrying is permitted. Inside any larger area one may not carry. The Mishnah[1] then starts with a dispute between R. Judah ben Baba and R. Akiba (early 2nd century) regarding the length of the side of a square of area 5,000 square cubits. R. Judah says that if the side is 70 and a fraction, one may carry within such a square only under certain conditions, whilst R. Akiba maintains that one may carry therein unconditionally. Obviously, R. Akiba accepted the value of the fraction as $\frac{2}{3}$. This would make the area of the square $(70\frac{2}{3})^2 = 4993\frac{7}{9}$, which is well within the 5,000 allowed. R. Judah, however, realised that the value $\frac{2}{3}$ (for the fraction) was a little too small and must have taken it as the next higher simple fraction $\frac{5}{7}$. By so doing one gets the area of the square as equal to $(70\frac{5}{7})^2 = 5000\frac{25}{49}$ which is just outside the permitted limit. It is interesting to note that $\sqrt{5000} > 70\frac{5}{7}$ gives $50\sqrt{2} > \frac{495}{7}$ i.e. $\sqrt{2} > \frac{99}{70}$.

[1] Erubin ii. 5, (23a.)

($\frac{99}{70}$ is an approximation given by Archimedes.)

The following method given in the Palestinian Talmud[1] of proving that $\sqrt{5000} > 70\frac{2}{3}$ is very instructive, as it makes use of geometrical figures for the purpose of Arithmetical calculations. " R. Samuel bar Nachman said in the name of R. Jonathan that the side of a square whose area equals that of a court 100×50 sq. cubits is more than $70\frac{2}{3}$, since a square of sides $70\frac{2}{3} = 70 \times 70$ (which $= 4900 = 5000 - 100$) $+ 70 \times \frac{2}{3} + 70 \times \frac{2}{3}$ (which $= \frac{140}{3} + \frac{140}{3} = \frac{280}{3}$) $+ 4$ squares of $\frac{1}{3}$ $(= \frac{4}{9})$ in each corner of the figure, and there is still lacking $\frac{19}{3}$ less $\frac{1}{9}$ "

If we look at the Fig. 24 where ABCD is a square whose sides are 70×70 and EFGH a square $70\frac{2}{3} \times 70\frac{2}{3}$ we see that EFGH = ABCD + 4 rectangles AK, BL, AM, DN, and the 4 corner squares AE, BF, CG, DH.

i.e., $\text{EFGH} = 70 \times 70 + 4(70 \times \frac{1}{3}) + \frac{4}{9}$

$= 4900 + \frac{280}{3} + \frac{4}{9}$

$= 4900 + 93\frac{1}{3} + \frac{4}{9} = 4993\frac{7}{9}$ square cubits.

The difference between this value and $5000 = 6\frac{2}{9} = \frac{56}{9}$

$$= \frac{57-1}{9} = \frac{57}{9} - \frac{1}{9} = \frac{19}{3} - \frac{1}{9}$$

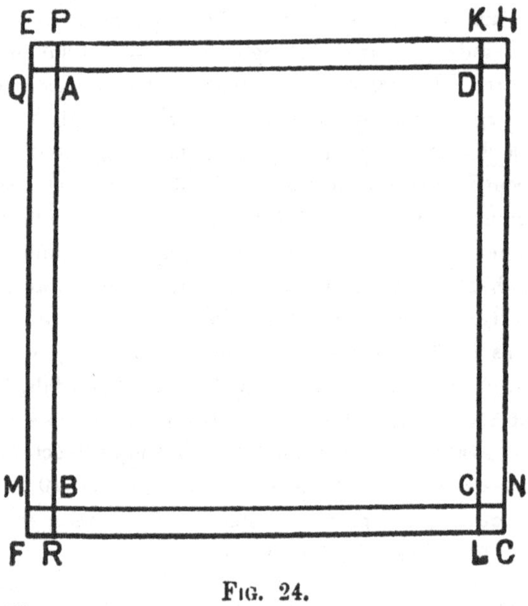

Fig. 24.

[1] J. Erubin 20b,

NOTE. Both the Babylonian and Palestinian Talmuds realise that the value of the fraction is slightly more that $\frac{2}{3}$, but whilst the Babylonian Talmud says that the sages *have not* determined the value more exactly[1] (ר' יהודה אומר דבר מועט יש על אמה ושיריים ולא נתנו חכמים בו שיעור) the Palestine tractate says[2] that they *could not* determine it more exactly (ולא יכלו חכמים לעמוד עליו), indicating the irrational nature of $\sqrt{5000}$.

Rashi to Erubin 23b. gives the following ingenious construction to prove the same thing (i.e., that $\sqrt{5000} > 70\frac{2}{3}$).

He says (without giving a diagram)" Let ABCD (Fig. 25) be a rectangle whose sides are 100 and 50 cubits respectively.

By joining the middle points EF of the sides AB, DC we get 2 squares 50 × 50 cubits. Divide AE, DF, into 5 equal parts Aa, ab, etc., Da′, a′b′, etc., then we get 5 rectangles of 50 × 50 cubits.

Attach one of these rectangles to each side of the square EFCB and you get a figure whose area is less than 70^2 by 4 times the area of each of the dotted corner squares, i.e., the area of the figures thus obtained = square RSVT less the 4 dotted squares = $70^2 - (4 \times 10^2)$.

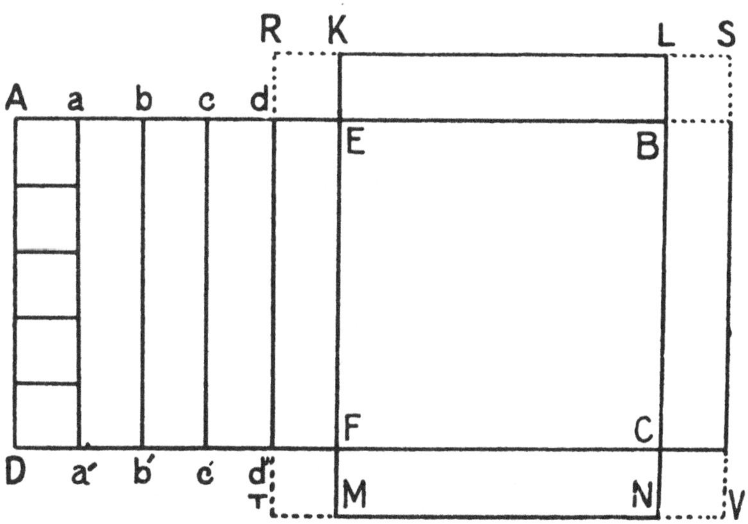

FIG. 25.

By further dividing the remaining rectangle Aaa′D into 5

[1] Erubin 23b. [2] Jer. Erubin 20b.

squares 10×10 and adding one to each of the 4 dotted corners of the figure so as to complete the square RSVT, we get a complete square 70×70.

Therefore the area 100×50 is greater than 70 by one of the squares 10×10.

By further dividing the remaining square of 10×10 into 30 rectangular strips of sides 10 and $\frac{10}{30}$ respectively and forming, out of those strips, one long rectangular strip of $300 \times \frac{10}{30}$ (i.e. 100 square cubits) and then attaching a portion of that rectangle of length 70 (and of area $\frac{70}{3}$) to each of the sides of the square RSVT, a new figure will be formed whose area will be less than $(70\frac{2}{3})^2$ by 4 times the square of $\frac{10}{30}$ (i.e., 4 times the square of $\frac{1}{3}$). These 4 corner squares can again be filled in to form a square $70\frac{2}{3}$ by $70\frac{2}{3}$ from the remaining portion of the rectangle (which $= 20 \times \frac{10}{30}$) and we shall still have a portion of it left. This process can be continued longer, but if we stop here we come to the following conclusion: "That the area of a rectangle 100×50 is a little more than that of a square $70\frac{2}{3} \times 70\frac{2}{3}$, i.e., $\sqrt{5000} > 70.66$ and that its value cannot be calculated exactly.

[It is, however, most remarkable that although Rashi displayed great genius in Mathematical calculation, he was quite ignorant of the most elementary mathematical facts. He was not aware that the sum of two sides of a triangle is greater than the 3rd, for he says: "If a ladder is to be placed 4 spans from the foot of a wall 10 spans high so as to reach its top, then the ladder must be 14 spans high (i.e., the sum of the two lengths), which of course is absurdly incorrect, the real minimum length of the ladder must be only $\sqrt{4^2+10^2} = \sqrt{116} = 10.7$ spans.

Tosefoth in loco points out Rashi's error and says rightly that a ladder 14 spans high may be placed even 10 ft. from a wall 10 spans high since it forms the diagonal of a square 10 by 10, which is 14 spans.

Rashi is further wrong in his statement to Erubin 76a that it is impossible to inscribe a square 4×4 in a circle whose circumference is less than 24, and he gives a lengthy explanation of his statement which is altogether erroneous.]

Maimonides says that "the side of a square having an area 5,000 square cubits cannot be found; in the same way as the true value of the relation between the circumference and the diameter of a circle cannot be found; " and he adds, " this is not due to our ignorance but to the peculiar character of the calculation"

(ואין זה לחסרון דעתנו אלא מפני טבע זה החשבון)

He says that " the square root of 5,000 is approximately $70\frac{5}{7}$, which is, however, slightly too high, since $(70\frac{5}{7})^2 = 5000\frac{1}{2}$ approx."

With this value of $\sqrt{5000}$, we get $\sqrt{2} = \frac{70\frac{5}{7}}{50} = \frac{495}{350} = \frac{99}{70} = 1.414$ which is correct to 3 decimal places.

$\sqrt{5}$. Tosefoth remarks[1] that although the diagonal of a square $1 \times 1 = 1{\cdot}4$, the diagonal of a rectangle 1×2 is a little less than $2{\cdot}4$, i.e., $\sqrt{5} < 2.4$ (real value $= 2.23..$).

Further, the Mishnah[2] observes that $\sqrt{12500} = 111$ and a fraction, or less than 112

i.e. $50\sqrt{5} > 111$

Therefore $\sqrt{5} > 2.22$

(This Mishnah is quoted by Eduard Mahler,[3] but I have been unable to find that statement.)

[1] Sabbath 85b. [2] Erubin 36b.

[3] " Zur talmudischen Mathematik " Zeitschr. f. Mathem. 1884, 1886.

PART III.

ASTRONOMY.

השמים מספרים כבוד אל ומעשה ידיו מגיד הרקיע (תהלים י"ט).

The heavens declare the glory of God, and the firmament sheweth His handiwork. Psalms, xix. 1.

CHAPTER VII.

Astronomical Knowledge.

הידעת חקות שמים (איוב ל״ח, ל״ג)

'Knowest thou the Ordinances of Heaven?'. Job, xxxviii, 33.

As the reader is probably unfamiliar with the nature of many of the astronomical phenomena, and especially with the terminology used in describing them, it is necessary to give here a brief account of some of the more important facts, as well as of the nomenclature used both in modern and ancient (Ptolemaic) astronomy.

A considerable number of matters of astronomical interest are mentioned in the Talmud, and several numerical results are given, but nowhere is there any indication of the manner in which these results are obtained. As examples, with all of which we shall deal presently, we may take the estimated duration of the synodic month, the length of the tropical year as given by Samuel and the most remarkably accurate one attributed to Samuel's contemporary R. Adda bar Ahaba, the duration of the different kinds of twilight, etc But the most interesting and most complicated of the problems which engaged the attention of the Rabbis was the calculation of the moment when the sun and moon would be in conjunction (i.e., when the centres of the sun, moon and earth would be in the same straight line — in the order named), and the time that the new moon would be visible, for although a new month was proclaimed only after reliable evidence had been received that the new moon had actully been seen, such evidence was checked by mathematical calculation.

בית דין מחשבין בחשבונות כדרך האיצטיגנינים, שיודעין מקומות הכוכבים ומהלכם, וחוקרים ומדקדקין עד שידעו אם אפשר שיראה הירח בזמנו... אם באו עדים ודרשום וחקרום כהלכה ונאמנו דבריהם, מקדשין אותו.[1]

In what follows we shall apply the methods of Modern Mathematics to the elucidation of these matters.

[1] Maimonides Kid. Hach. i. 6.

The Celestial Sphere is a sphere whose centre is the centre of the earth, whose radius is of any arbitrary length, and upon whose surface are projected the various heavenly bodies in the directions in which they would be seen by an imaginary observer at the centre —instead of at the surface of the earth. Such a sphere is, of course, a purely mathematical conception and can have no real existence, since not only can no observer be situated at its centre, but also we know that the various celestial bodies are, in fact, at different distances from the earth, and cannot, therefore, lie on the surface of a sphere. The ancients, however, really believed in the existence of such a solid crystalline sphere, to the surface of which the stars were attached. The Talmud[1] refers to such a celestial sphere as *Galgal* (גלגל).

Parallax.

As astronomical observations of the positions of heavenly bodies are, of course, made from the surface of the earth, those positions as seen by the observer are not the same as would be seen by the observer at the centre of the earth. Thus, a body situated at S (Fig. 26) would appear to an observer on the surface of the earth to be at S_1 on the celestial sphere, although its true position on the sphere — as would be seen from the earth's centre — is at S_2. This angular difference ASC (or arc S_1S_2) between the true and apparent positions of the body is called its *Parallax* (שינוי מראה)[2] and its effect is to make the body appear lower in the heavens (S_1) than it really is (S_2). It is only when the body is at the *zenith* (Z) i.e., at the point on the celestial sphere vertically above the observer, that parallax is zero, and the true position coincides with the apparent position. On the other hand, when it is on the *horizon* (אופק) — say at S′, parallax (i.e., angle AS′C) is greatest.

(אצל האופק... שינוי מראה הוא עומד בתכלית גדלו).[3]

In the case of distant bodies like the fixed stars or sun, parallax is so small as to be negligible. Thus even in the case of the sun, horizontal parallax AS′C is an angle whose Sine is:

$$\frac{\text{AC (i.e., radius of earth)}}{\text{CS}' \text{ (i.e., distance of sun)}} = \frac{4{,}000 \text{ miles}}{93{,}000{,}000 \text{ miles}} = 0{\cdot}0004$$

i.e., AS′C = about 8″, which is negligible. In the case of stars

[1] Pesachim 94b. [2] Yesod Olam iii. [3] Ibid.

whose distances are still greater, $\frac{AC}{CS'}$ and therefore, parallax is still less. In the case of the moon M, however, whose distance is only about 60 times the radius of the earth $\frac{AC}{MC} = \frac{1}{60} = 0{\cdot}017$, and, therefore, AM′C is about 1°. In other words, when the moon appears to be on the horizon, it is actually already about 1° above the horizon in the celestial sphere, and hence parallax must be taken into account. The Talmud does not refer to parallax, but as the Rabbis had to check the evidence of witnesses (regarding the moon's visibility) by mathematical means they must have known how to correct for it. We shall see in a later chapter how Maimonides deals with the matter.

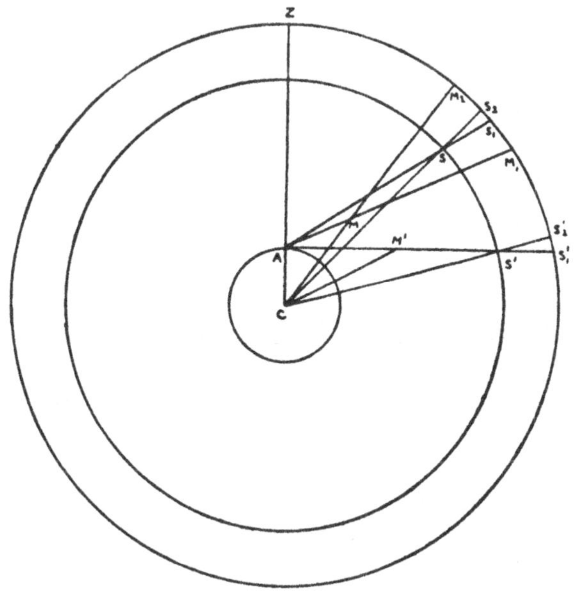

FIG. 26.

The Diurnal Rotation of the Heavens. To an observer watching the sky on a clear night, the heavenly bodies appear to be moving from East to West in arcs of parallel circles. (Fig. 27). In the course of the whole night some of the stars are seen to rise at the eastern horizon and some to sink at the western horizon (e.g. those on circles BAB′, EQW and GFG′). Others (like NK) make smaller circuits in the sky, such that they never, either at night or by day, sink below the horizon and can, therefore, (in the course of the 24 hours), be seen throughout their whole course. The Common Diameter or Axis PP′ around which this diurnal rotation of the heavens (מסיבה היומית) takes place is called the Celestial Axis (קוטר ; בריח) and each extremity of this diameter on the surface of the sphere is called a Pole (קוטב ; ציר). There is a North Pole, P, (ציר צפוני) and a South Pole, P′ (ציר דרומי). All stars lying between the North Pole and the Circle KN (which are never seen to *set* below the horizon in the Northern hemisphere) are called *Circumpolar Stars.* The Circumpolar Stars of the Southern hemisphere (i.e., those lying between the South pole and the circle SS′)

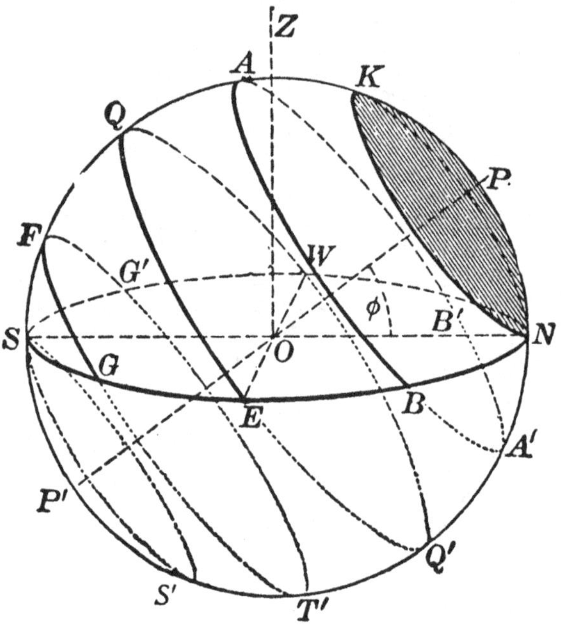

FIG. 27.

are never seen to *rise* in the Northern Hemisphere.

Definitions. The following points and circles on the celestial sphere are important (see Figs. 27 and 28):

The *Axis* and *Poles* have already been defined.

The *Zenith*, Z (נקודת נכח הראש) and *Nadir*, N (נקודת נכח הרגלים) are the points vertically above and below the observer respectively.

While the poles are fixed points in the sphere, the zenith and nadir necessarily vary with the position of the observer.

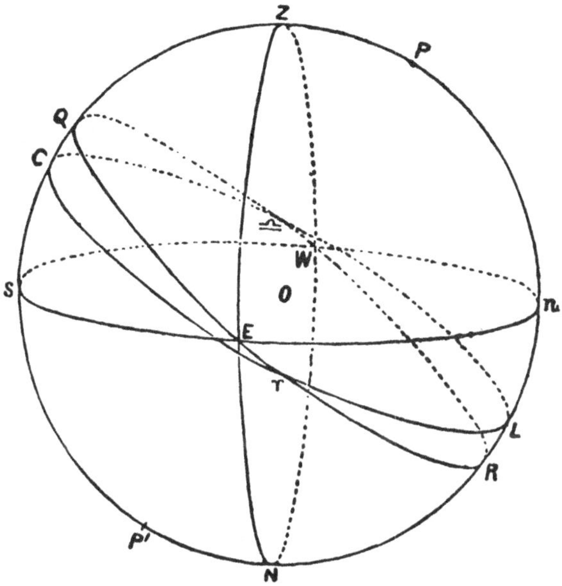

Fig. 28.

The Celestial Equator, QQ' in Fig 27 (עיגול המשוה), is the Circle which lies midway between the poles, or, which is the same thing, it is the plane in which the earth's equator cuts the celestial sphere. The diurnal circles of rotation of the stars etc. are parallel to the equator.

The Celestial Horizon, SEBN (אופק), is the circle which lies midway between the zenith and the Nadir. This, which passes through the centre of the earth, is the *True Horizon* (אופק האמתי), and is parallel to, but is not the same plane as, the *Sensible Horizon* AM'S' in Fig. 26 (אופק הנראה), which passes through the observer on the surface of the earth. The distance between the two horizons is, of course, equal to the radius of the earth.

The Meridian is the circle which passes through the zenith and nadir as well as the two celestial poles, ZPNP' (Fig. 28). It cuts the horizon at the North and South points, N and S; the equator and horizon intersecting at the *East* and *West* points, E and W. From Fig. 27 it is seen that the Meridian bisects all the diurnal circles, so that in the case of the stars that rise and set (e.g., B and G), when they cross the meridian at A and F they are just midway between the time of rising and setting. The same of course, applies to the sun so that when the sun is on the meridian it is just midway between sunrise and sunset, i.e. it is local noon. Hence the name *Meridian* (אופן הצהריים). The figure also shows that at the moment any star has reached the meridian in its course from east to west its *altitude* (גובה), or angular height above the horizon is highest, and when it is on the meridian in its course from west to east, its altitude is lowest. Hence the direction of the meridian at any place can be found by finding the point in the heavens at which the sun has the maximum altitude and drawing a vertical plane through that point and the zenith (see p 84).

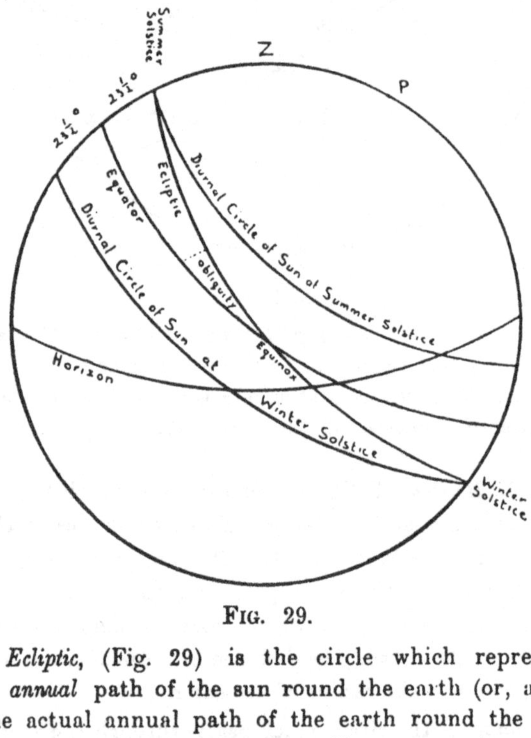

FIG. 29.

The Ecliptic, (Fig. 29) is the circle which represents the apparent *annual* path of the sun round the earth (or, as we now know, the actual annual path of the earth round the sun). It derives its name from the fact that eclipses can only occur when the moon is at or near the plane of this circle (see Eclipses). It is therefore called in Hebrew גלגל הקדרות. Whilst the sun's diurnal circle is like the diurnal circle of any other heavenly body parallel to the equator, and the direction of motion is from East to West, the plane of the ecliptic is inclined to that of the equator at an angle called the *obliquity of the ecliptic* (נטייה), whose present value (1929) is 23°26′54.67″. Its value was ascertained with remarkable precision by the Chinese about 1100 B.C. They found it to be 23°52′. It is probable that the Chaldeans and Egyptians knew its value with considerable accuracy long before them. Hipparchus (2nd century B.C.) and Ptolemy (2nd century C.E.) gave its value as 23°52′, but the Talmud nowhere gives an estimate of its value. It is certain, however, that the Talmudic Rabbis knew its size since, as we shall see, they knew the principle of the Gnomon by means of which the Ancients measured this angle and obtained other Astronomical Constants (see p. 83). Observations by different observers from the time of Ptolemy onwards—e.g. the Arabians, Al Battani (about 900), Zarkali (about 1050), and the Jewish physician and astronomer Prophatius (1303)—show a gradual and progressive diminution in the value of this angle. Maimonides[1] gives its value as about 23½°. It has been shown on dynamical principles that there is a limit to this diminution, i.e., the ecliptic will never coincide with the equator, but that the angle fluctuates in size, in the course of about 20,000 years, between 24°36′, and 22°15′., i.e., at the rate of about 0·47″ a year. As its present value (1929) is 23°26′54.67″ therefore its value at the time of Maimonides (1178) must have been $23°26'54{\cdot}67'' + (1929-1178) \times 0{\cdot}47'' = 23°32'17{\cdot}64''$ which is about the same as that given by him. Israeli[2] gives its value as "a little under 24°" which is a very rough estimate.

In addition to the *Secular diminution* of the obliquity at the rate of 0.47″ per year the angle undergoes another variation due to *Nutation* (see p. 77).

Great and Small Circles. Every circle on the surface of a sphere whose plane passes through the centre (and thus bisects the sphere) is called a *Great Circle* (עגול גדול) because no greater circle can be drawn on it. A *Small Circle* (עגול קטן) is one whose plane does not pass through the centre of the sphere.

The Horizon, Meridian, Equator and Ecliptic are great circles, whilst the circles of diurnal rotation of the stars (other than those in the equator) are small circles (see Fig. 27). It is to be noted that as any two great circles intersect in a common diameter, they necessarily bisect one another. Hence the 4 points of the Compass (the *Cardinal Points*) divide the horizon into 4 equal parts of 90°

[1] Kiddush Hachodesh xix. 6. [2] Yesod Olam ii. 5.

each. The East and West points are the midpoints of the Eastern and Western Horizons respectively (חצי מזרח, חצי מערב)[1], whilst the North and South points are the midpoints of the Northern and Southern horizons respectively.

R. Josi gives the following correct *method of finding the Cardinal Points*: East and West are the points on the horizon where the sun rises at the two equinoxes (i.e., on the days when the sun is on the equator תקופת ניסן ותקופת תשרי). Where the sun rises or sets in the long Summer days, is north of the East or West points; where it rises or sets in the short Winter days is South of these points.[2]

אמר ר״ יוסי...חמה יוצאה ביום ארוך ושוקעה ביום ארוך זה הוא פני צפון, חמה יוצאה ביום קצר ושוקעת ביום קצר זה הוא פני דרום, תקופת ניסן ותקופת תשרי חמה יוצאה בחצי מורח ושוקעה בחצי מערב.

The Ptolemaic and Copernican Systems. On p. 65 I spoke of the diurnal rotation of the heavens. We know now that this movement is apparent only, and is due—as Copernicus conclusively showed in 1543 C.E.—to the daily rotation of the spherical earth upon its axis from West to East. Until the time of Copernicus, however, although some of the earliest Greek Astronomers, like Pythagoras (6th century B.C.), believed that night and day were due to the spinning of the earth round its axis, and the seasons were caused by the revolution of the earth round a fixed sun (helio-centric system), nevertheless, the views of Ptolemy (2nd century C.E) prevailed that it was the stars and other heavenly bodies that moved round a fixed earth (Geocentric system).

Talmudic View. The view of the Talmudic Rabbis on the matter is open to some controversy, which centres round the following passage[3] :–

תנו רבנן חכמי ישראל אומרים גלגל קבוע ומזלות חוזרין, וחכמי אומות העולם אומרים גלגל חוזר ומזלות קבועין.

"The Rabbis taught: 'The Jewish sages say that the *Galgal* is fixed and the *Mazoloth* rotate, whilst the Gentile sages say that the *Galgal* rotates and the *Mazoloth* are fixed.' "

Those who are imbued with the idea that the Rabbis' views are always in agreement with the most up-to-date scientific discoveries, suggest that Galgal, which means a circle or sphere, refers to the

[1] Erubin 56a. [2] Erubin 56a. [3] Pesachim 94b.

sun and Mazoloth refers to the planets. The passage would, therefore, tend to show that the Jewish Sages were followers of the Pythagorean school and anticipated the Copernican heliocentric system. Whilst those who believe that the Jews were upholders of the Ptolemaic geocentric system, would make Galgal refer to the earth. As a matter of fact, it is quite certain that both of these interpretations are incorrect. It is true that *Galgal* may refer to the sun, but *Mazoloth* – although it sometimes refers to the planets[1]– cannot in this instance have that meaning, since the passage makes it quite clear that it refers to fixed stars, such as the Waggon or Great Bear (עגלה) and the signs of the Zodiac, such as the Scorpion (עקרב). On the other hand the Jewish view cannot be in accord with the Ptolemaic system of a fixed earth, since Galgal nowhere refers to the earth. The true interpretation of the passage is almost certainly the one given by Rabbi Abraham, son of Maimonides, who makes Galgal refer to the celestial sphere. The view of the Gentile sages is thus in agreement with the Ptolemaic system, viz., that the stars are fixed on the surface of the celestial sphere, which latter moves round the earth carrying the stars with it. Indeed R. Achi bar Jacob explicitly states in the same passage that the *Galgal* and *Mazoloth* rotate together as one rigid body —" like a millstone round an axle " (כבוצינא דריחיא), or " like a door round its hinges " (כצינורא דדשא) The Jewish view would then be somewhat fantastic and unintelligible, and is made still more fantastic by the continuation of the passage:

חכמי ישראל אומרים ביום חמה מהלכת למטה מהרקיע ובלילה למעלה מן הרקיע וחכמי אומות העולם אומרים ביום חמה מהלכת למטה מן הרקיע ובלילה למטה מן הקרקע.

" The Jewish sages say that during the day the sun moves below the sky, and at night above the sky (and cannot therefore be seen), whilst the Gentile sages say that it moves below the sky in the day and below the earth at night. ' But the final words of the passage make it quite clear that, after full discussion, *the Rabbis adopted the Ptolemaic system*, for they conclude

א"ר ונראין דבריהן מדברינו.

" Said an anonymous Rabbi : ' The Gentile view seems more probable than ours.' "

[1] E.g. Sabbath 156a, אין מזל לישראל which means that the Jews must not believe in the influence of the planets upon the careers of man.

It is interesting to note the reason which prompted the Rabbis to adopt the Ptolemaic system, viz. that " during the day the pools are cold (from which it is inferred that the sun is high above the earth) and at night they are warm (because the sun is not far under the earth) " שביום מעינות צוננין, ובלילה רותחין.

Such an explanation is, of course, entirely erroneous, because not only does it assume that the earth is *eccentrically* situated inside the celestial sphere, but that the eccentricity of its position, i.e., the distance between its centre and the centre of the celestial sphere is very considerable. As a matter of fact, it is agreed by all that the apparent diurnal rotation of the heavens (including that of the sun) is round the earth as centre. It would seem that the Rabbis confused the sun's diurnal rotation round the earth as centre with its annual path, of which—as we shall see—the earth is not the centre, but whose eccentricity as we now know is no more than about $\frac{1}{60}$ (see p. 110).

The diurnal rotation of the heavens is also mentioned in an anecdote related by the famous Talmudic fabulist Rabbah Bar Bar Channah [1] He once went to a place where earth and heaven meet (דנשקא ארעא ורקיעא אהדדי) and he found some holes in the sky, into one of which he placed his food basket which soon disappeared. In reply to his enquiry as to whether there were any thieves there he was told that ' the heavens revolve round the earth and if you wait till to-morrow and you will get it back.'

האי גלגלא דרקיעא הוא דהדר, נטר עד למחר הכא ומשכחת לה.

The Ptolemaic system, as we saw, held sway till the 16th century (when Copernicus in 1543 published definite proof of the heliocentric system) and formed the basis of all the writings of the Arabian and Jewish Astronomers of the Middle Ages—including Maimonides and Abraham bar Chiya of the 12th century, Isaac ben Joseph Israeli[2] and Levi ben Gershon of the 14th century, and others. The Copernican system met with very determined opposition, and the Christian Church as well as the Jewish Rabbis denounced it as heretical because the Bible said that the earth was fixed and the sun was moving. Galileo was thrown into prison for

[1] Baba Bathra 74a.

[2] Not to be confused with Isaac ben Solomon Israeli (Isaac Judaeus), a prominent physician of the 10th and 11th centuries.

his unorthodox views, and it is therefore not surprising that Rabbi Tobia Katz, physician and cyclopaedist of the 17th century, wrote vehemently against the Copernican theory[1] : " It behoves all theological philosophers to reject the Copernican system because it is contrary to the teachings of holy Scripture and the words of the true prophets whose statements are correct—and they state plainly (Eccles. i. 3 and 4) that ' the earth standeth for ever; The sun ariseth, and the sun goeth down and hasteth to his place where he rose.' " Even at the present day some of the ultra-orthodox Rabbis refuse to accept the Copernican theory, although Midrashic Authority, were such necessary, can be produced in its favour, for Genesis R. 5. says that " the earth is called *eretz* (ארץ) because it runs (רץ) to do the will of God."

Music of the Spheres. All the ancient and mediaeval astronomers from the time of Pythagoras onwards up to the time of Kepler (17th century), taught that the various heavenly bodies, in their movements round the earth, emitted musical sounds, which, specially gifted persons are, at times, privileged to hear. Job (xxxvi. 7) speaks of the morning stars singing in harmony (ברן יחד כוכבי בוקר). The Psalmist refers to it poetically a couple of times, (xix. 2-4, and xlii. 9), whilst the Talmud[2] speaks about the noise made by the sun in its rotation through space—because it cuts its way through, like a saw cutting through timber, and produces the sawdust which is seen in a strong beam of light

תנו רבנן, שלשה קולות הולכין מסוף העולם ועד סופו, ואלו הן קול גלגל חמה, "וכו׳".

Shakespeare refers to it a number of times, e.g. in The Merchant of Venice, Act 5, scene 1:

" There's not the smallest orb which thou behold'st
But in his motions like an angel sings."

Equinoxes and Solstices (תקופות). At each point in the ecliptic the sun takes part in the diurnal rotation of the heavens (Figs. 29 and 30). Hence at each of the two points at which the ecliptic crosses the equator (נקודות הנגיעה) the sun's diurnal circle coincides with that of the equator, and half of that circle lies above the horizon and half of it below it. The result is that on those two occasions day night are equal throughout the world. For this reason

[1] Masse Tobia (Olam Hagalgalim iii). [2] Yoma 20b.

the two points of intersection of the ecliptic with the equator are called the *equinoctial points*, or *equinoxes* (השואות). The *Vernal* or *Spring Equinox* (השואת הקיין) now occurs on the 21st of March, and the *Autumnal Equinox* (השואת החורף) now occurs about the 23rd September. After each equinox the sun gradually recedes from the equator until it reaches the greatest angular distance of about 23½° (i.e., the obliquity of the ecliptic) at S and S^1 respectively, when the sun changes its direction with respect to the equator and instead of receding from it begins to approach it again. At the points S and S^1 therefore the sun appears to be standing still, momentarily, and for this reason these points are called the *Solstices* (עמידת השמש). Israeli calls these points הפוכים or *turning points*, or *tropical points*. The *Summer Solstice* (הפוך החום), now occurs about 21st June and the *Winter Solstice* (הפוך הקור) on the 22nd December. In the Talmud the 4 points are collectively called the T'kufoth (תקופות) and they mark the beginnings of the 4 seasons as follows: [1]

Sun at Vernal Equinox תקופת ניסן, Beginning of Spring, March 21.
Sun at Summer Solstice תקופת תמוז, Beginning of Summer, June 21.
Sun at Autumn Equinox תקופת תשרי, Beginning of Autumn, Sep. 23.
Sun at Winter Solstice תקופת טבת, Beginning of Winter, Dec. 22.

The T'kufoth of Samuel and of R. Adda. For reasons which will be explained later, the seasons, or t'kufoth, i.e., the intervals between an equinox and the following solstice or between the latter and the following equinox—are not constant. Thus at the present time their respective lengths are as follows: Spring = 92·84 days; Summer = 93·6 days; Autumn = 89·8 days and Winter = 89·02 days. The *average* interval, however is, of course, *one quarter* of a *tropical year*—the latter being the interval between two consecutive vernal equinoxes. Samuel [2] made the average interval between an equinox and a solstice equal to 91 days 7½ hours:

אמר שמואל, אין בין תקופה לתקופה אלא תשעים ואחד יום, ושבע שעות ומחצה.

This interval is known as Samuel's t'kufah תקופת מר שמואל and it makes the tropical year 91¼ × 4 = 365 days, corresponding to a Julian year.

[1] The term t'kufah is also applied to the season itself. Thus t'kufath Tebeth refers not only to the winter solstice but also to Winter itself (Rosh Hashanah 21a).

[2] Erubin 56a.

A more accurate value is, as we shall see later, attributed to R. Adda bar Ahaba (a contemporary of Samuel), viz. 91d. 7h. 28m. 51·34s. This is known as t'kufath R. Adda תקופת ר" אדא and makes the tropical year equal to 365d. 5h. 55m. 25·4385s., a value which differs by about 12½ secs. from Hipparchus' value of 365d. 5h. 55m. 12s. This t'kufah is nowhere specifically mentioned in the Talmud, nor does Hai Gain of the 10th Century refer to it. For this reason Slonimsky and others are of opinion that not only were the Talmudic Rabbis unaware of it, but that the t'kufah going by the name of R. Adda was not introduced before the 10th century —by some anonymous person. In support of this contention is also the fact that R. Adda bar Ahaba is, with the exception of Erachin 9b, nowhere, in the Talmud, mentioned in connection with Astronomical matters at all, as is the case with Samuel. Piniles,[1] however, maintains that R. Yochanan of the 3rd century made use of this t'kufah in his capacity of member of the Calendar Council. His evidence, which, though not conclusive, is very ingenious, is based upon the following passage in the Palestinian Talmud.[2]

רב אמר סטורנלייא לפני תקופה ח' ימים... ר" יוחנן פרוקטו ראשה של תקופה.

" Rab said the saturnalia (which began on the 17th December) are 8 days before the t'kufah (of Tebeth). R. Yochanan said, the '*Procto*' is the beginning of the t'kufah." Piniles believes that Procto is the Greek word πρόχθες which means " the day before yesterday " and therefore urges that according to R. Yochanan the Saturnalia began 2 days before the t'kufah. Hence while Rab calculated the date of t'kufath Tebeth (of the year 4009 of the world era or 248 B.C.) to take place on the 25th December—which would be the correct date for that year according to Samuel's t'kufah, R. Yochanan made it occur on the 19th December—which as will be shown, is in agreement with R. Addas' t'kufah. Piniles dismisses the suggeston of R. Moses Margolioth of the 18th century (פני משה) that *Procto* is a corruption of the Greek word Tropic, i.e., " turning-point " (which would in this case be the Solstice) by the plausible statement that there is no reason for R. Yochanan to point out—in connection with the Roman Festivals– that the Solstice is the beginning of the t'kufah, since that event is

[1] דרכה של תורה Wien 1861 pp. 147-150. [2] J. Abodah Zarah i. 2.

not a festival. On the other hand it seems to the author that if the suggestion of Piniles is correct the sentence should read ..בארוקטו של ראש התקופה (the 2nd day before the beginning of the t'kufah)

Precession of the Equinoxes (העתקה). If one measures the time of a complete revolution (360°) of the sun from a given fixed star to the same star again (i.e., *sidereal year*), and one compares this interval with the time it takes the sun to travel from vernal equinox to vernal equinox (*tropical year*), one finds that the latter is less than the former by about 20⅓ minutes. As the sun covers 360° in about 365¼ days, its rate of movement is about 0°.59′.8″ per day, or about 148″ per hour. Hence, in 20⅓ minutes the sun covers 50·1″ of arc. It follows, therefore, that between two consecutive vernal equinoxes the sun has to travel 50·1″ of arc less than 360°. The only possible explanation of this is that the equinoctial point must itself be moving on the ecliptic to meet the sun (i.e., from east to west). This retrograde motion of the equinoctial point to meet the sun is called **precession** of the equinoxes — because it makes the equinoxes precede their due time every year, and its rate is 50·1″ a year (Fig. 30).

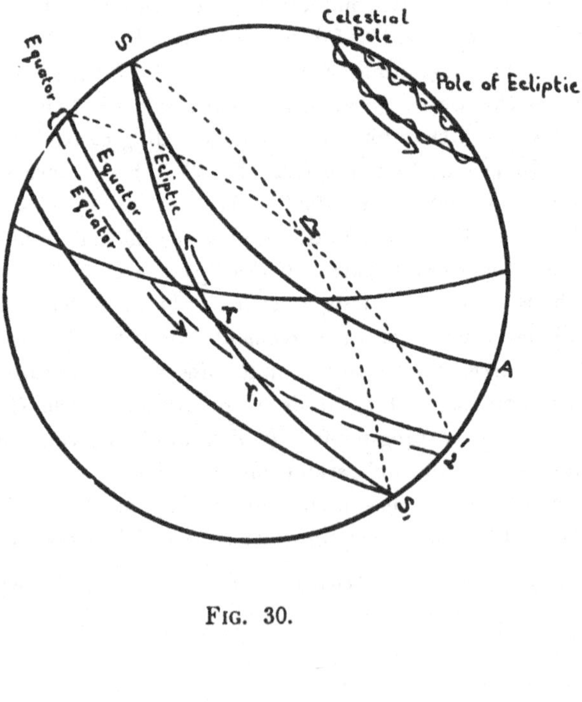

FIG. 30.

Precession was discovered by Hipparchus, and was possibly known to the Babylonians, but is not mentioned in the Talmud. Its occurrence can be explained by Newton's law of Gravitation. When the vernal equinox was first located in the heavens by Hipparchus it was at the beginning of the sign of the zodiac called the Ram or Aries (1st point of Aries). Since then it has moved 30° to the sign of Pisces, but is still called the First point of Aries, and its symbol is ♈. Hence precession may be called in Hebrew תנועת הטלה (the movement of Aries). Ptolemy made its rate equal to 1° in 100 years (see Yesod Olam ii. 1). The Babylonians, according to Mahler,[1] made it equal to 1° in 72 years. Its value is now known to increase at the rate of 0·000,244,297″ per year—an amount of arc which the sun covers in $\frac{1}{1000}$ of a second. Hence the tropical year is gradually decreasing at the rate of about $\frac{1}{1000}$ sec. per year.

NOTE: The rate of precession has been used to fix ancient dates. Thus, according to some, the Pleiades (כימה), which is in Taurus, was in Job's time, the Spring Equinox (Job xxxviii). If that is so the date of Job would be 2170 B.C.

A Precessional Year is the period of a complete rotation of of the equinoctial point round the ecliptic. As the rate of precession is 1° in 72 years, therefore a precessional year $= 72 \times 360 =$ about 25,920 years (say 26,000 years).

Cause of Precession. Precession, which, as I said, can be explained by the law of gravitation, is brought about by the rotation of the celestial pole (i.e., the pole of the equator) in the form of a small circle round the pole of the ecliptic, thus making the equator shift round the ecliptic. The rotation of the celestial pole does not, however, occur in the form of a smooth circumference of a circle, but as a wavy path (Fig. 30), the mean position of which lies on the circumference of a small circle. This nodding or *Nutation* of the celestial pole (a phenomenon discovered by the English astronomer Bradley) necessarily causes a corresponding nodding of the equator, thus bringing about a *fluctuation* or alternate increase and diminution *in the value of the obliquity*. The maximum amount of nutation is 9″ above or below the mean position of the celestial pole and therefore the value of the

[1] Handb. d. Jüd. Chronol. Leipzig 1916, p. 22.

obliquity must periodically fluctuate between $23°.27'.8'' + 0°.0'.9''$ and $23°.0'.8'' - 0°.0'.9''$. As it has been found that the period of such fluctuation is about $18\frac{2}{3}$ years, it follows that each nutation wave lasts about $18\frac{2}{3}$ years. But the complete rotation of the pole (i.e., the precessional year) lasts about 26,000 years. Therefore the total number of nutation waves in the circle is $\frac{26,000}{18\frac{2}{3}}$ = about 1,400, and therefore the length of each wave = $\frac{1400}{360°}$ = about $15'.24''$.

Trepidation. Up to the time of Tycho Brahe (16th century), it was believed on the authority of Tobit ben Korra (9th century) and Zarkali (11th century)—two eminent Arabian Astronomers—that the equinoctial point, instead of moving continually in the opposite direction to that of the sun, undergoes periodical changes in direction, so that for a fixed period of years it moves in one direction—say from east to west to meet the sun, thus making the tropical year shorter than the sidereal—and then it suddenly changes its direction—say from west to east, running after the sun, thus making the tropical year longer than the sidereal—for an equal period of years. This is called *trepidation*, or trembling, and Israeli[1] refers to it in the following words:-

אברהם הורקאלי וחבריו... הודו ואמרו כי... העתקה ונסיעה יש לגלגל המזלות...
אבל אינו כלפי פנים לעולם... אלא פעם וזמן רב ידועה היא העתקתו כלפי פנים
ועל סדר המזלות, ופעם אחרת וזמן רב אחר ידוע הוא חוזר ונעתק ונוסע כלפי אחור.

" Abraham Zarkali and his colleagues said that there is a movement of the ecliptic . . . but not always in a forward direction . . . But sometimes and for a fixed long period it moves forwards in the direction of the Signs of the Zodiac, and at another time, again for a fixed long period, the movement is in the opposite direction."

This theory, which prevailed between the 9th and 16th Centuries, is nowhere referred to by Maimonides, a fact which makes it probable that he did not believe in it.

The Zodiac (חגורת המזלות) is a belt of the heavens about 16° wide, viz., 8° on each side of the ecliptic, containing 12 *groups* of stars, fancifully supposed by the ancients to resemble figures of animals—with the exception of one, viz., Libra " the Scales "

[1] op. cit. ii. 6.

(המזלות האמתיים בעלי הצורות) This belt comprises the orbits of the sun moon and the known planets. The 12 Groups or Constellations—called signs (סמנים) —which divide the zodiac into equal arcs of 30° each have the following names and symbols—for the origin of which the reader is referred to the article on the History of Astronomy,by Miss Agnes Clarke,in the Encyclopaedia Britannica.

Names			Symbol	Longitude	Month
Latin	English	Hebrew			
1. Aries	Ram	טלה	♈	0–30°	Nisan
2. Taurus	Bull	שור	♉	30°–60°	Iyar
3. Gemini	Twins	תאומים	♊	60°–90°	Sivan
4. Cancer	Crab	סרטן	♋	90°–120°	Tammuz
5. Leo	Lion	אריה	♌	120°–150°	Ab
6. Virgo	Virgin	בתולה	♍	150°–180°	Elul
7. Libra	Scales	מאזנים	♎	180°–210°	Tishri
8. Scorpio	Scorpion	עקרב	♏	210°–240°	Mar Cheshvan
9. Sagittarius	Archer	קשת	♐	240°–270°	Kislev
10. Capricornus	Goat	גדי	♑	270°–300°	Tebeth
11. Aquarius	Waterman	דלי	♒	300°–330°	Shebat
12. Pisces	Fishes	דגים	♓	330°–360°	Adar

Rashi to Baba Metzia 106b gives the following mnemonic composed of the initials of the signs: TeSHeTH (ט"שת) S'AB (סא"ב) MaAK (מע"ק) G'DaD (גד"ד).[1]

[1] See also Maimonides Kidd. Hach. 1. 3. and Israeli op. cit. ii. 16.

CHAPTER VIII.

Astronomical measurements.

ועמידם לעד לעולם חק נתן ולא יעבור (תהלים קמ"ח.)

He hath established them for ever. He gave a decree which none shall transgress (Psalms clxviii)

Celestial Localisation. (Fig. 31) Supposing the positions of the ecliptic and of the equator as well as of their points of intersection (the equinoctial points) to be accurately known on the celestial sphere, then the position of any heavenly body B is completely determined if we know either of its following pairs of **Coordinates** :

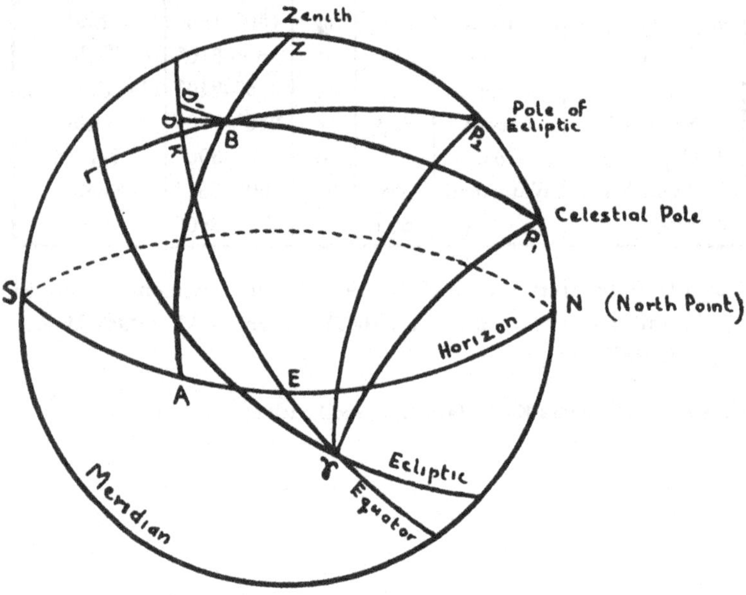

Fig. 31.

1. ECLIPTIC COORDINATES.

(a) Its *Latitude* (רוחב) i.e., its perpendicular angular distance BL from the ecliptic; and (b) Its *Longitude* (אורך), i.e., its angular distance ♈L from the vernal equinoctial point ♈ along the ecliptic.

2. EQUATOR COORDINATES.

(a) Its *Declination* (נטייה), i.e., its perpendicular angular distance BD from the equator; and (b) its *Right Ascension* (מצעד הישר or simply מצעד), i.e., its angular distance ♈ D from the vernal equinoctial point as measured along the equator.

As the equator and ecliptic are, but for minor alterations which we have considered under *precession*, fixed circles on the celestial sphere, these coordinates are sufficient to determine the position of a celestial body without reference to the position of the observer on the earth's surface. There is, however, still a 3rd pair of coordinates, which, owing to their constant variation on the celestial sphere, determine the position of a heavenly body at a given moment and for a given latitude on the earth's surface only. These coordinates are known as:

3. HORIZON COORDINATES.

(a) *Altitude* (גובה) i.e., perpendicular angular distance BA above the observer's horizon; and (b) *Azimuth* i.e., its angular distance SA from the South point.

Spherical Trigonometry. In order to be able to make use of these coordinates for the purpose of Astronomical computation, a knowledge of Spherical Trigonometry is necessary and the reader will, therefore, have to take the various formulae used in spherical trigonometry for granted.

A **Spherical Triangle** (Fig. 32) (משולש כדורי), is the figure on the surface of a sphere bounded by the arcs of 3 great circles (משולש העשוי מג' קשתות שמעגולות גדולות).[1] Thus in Fig. 32a the arcs AB,

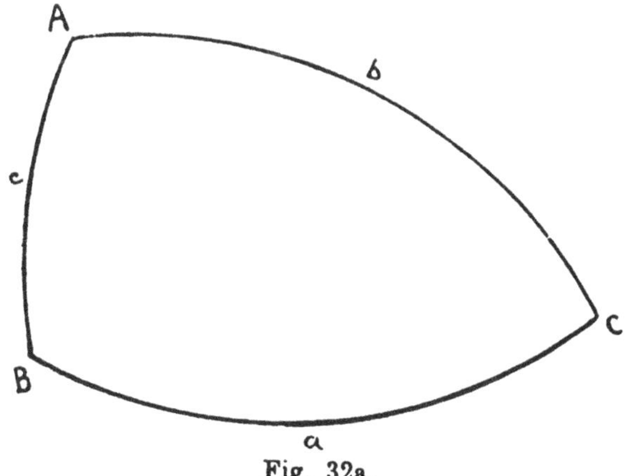

Fig. 32a.

[1] Yesod Olam i. 2.

BC, CA, are the sides of the triangle ABC and the angles BAC, ABC, BCA, are its angles. If we call the angles A, B, and C respectively, then their opposite sides are generally designated *a*, *b* and *c* respectively. When any 3 parts of a spherical triangle are known, the other 3 parts can easily be calculated (i.e., the triangle can be solved) by means of certain well known formulae connecting the known with the unknown parts. These formulae represent the *Properties* (סגולות) *of Spherical Triangles*, of which the following are important for our purpose, and are given here for reference.

(a) *Properties of all kinds of Spherical Triangles.* (ABC, Fig. 32a).

(i) $$\text{Cos A} = \frac{\text{Cos a} - \text{Cos b Cos c}}{\text{Sin b Sin c}} = \frac{\text{Cos a}}{\text{Sin b Sin c}} - \text{Cot b Cot c} \qquad \text{.......(1)}$$

This formula is useful for solving a triangle whose 3 sides are known. As a **corollary** thereof it follows that if $a = 90°$ (i.e., the side a is a quadrant) then:-

$$\text{Cos A} = -\text{Cot b Cot C (because Cos a} = \text{Cos } 90° = 0) \text{......(1a)}$$

(ii) $$\frac{\text{Sin a}}{\text{Sin A}} = \frac{\text{Sin b}}{\text{Sin B}} = \frac{\text{Sin c}}{\text{Sin C}} \text{........................(2)}$$

Israeli[1] calling the Sine of an angle בקע קשת, gives relation (2) as follows:

ערך בקע קשת צלע בג, אצל בקע קשת זוית א, הוא כערך בקע קשת צלע בג אצל בקע קשת זוית ב וכו'.

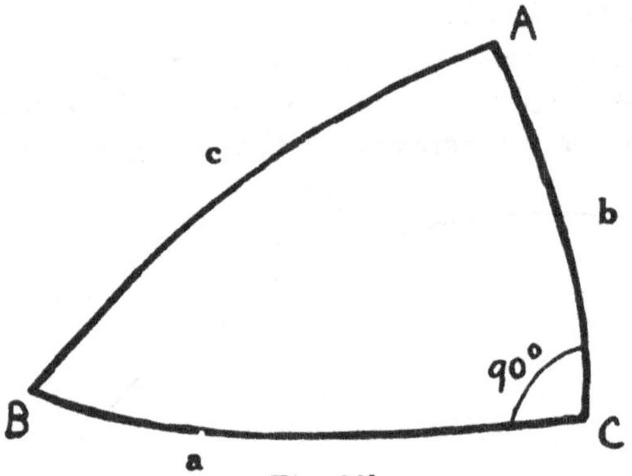

Fig. 32b

(b) *Properties of Right-Angled Spherical Triangles.* (Angle $C = 90°$). Fig. 32b.

(iii) $\text{Cos c} = \text{Cot A Cot B}$.................(3)

(iv) $\text{Sin B} = \dfrac{\text{Cos A}}{\text{Cos a}}$ (or $\text{Cos a} = \dfrac{\text{Cos A}}{\text{Sin B}}$)...(4)

} **Useful for solving right angled triangles whose other 2 angles are known.**

[1] Yesod Olam i. 2.

Israeli[1] calling the cosine (i.e., the Sine of the complement) of an angle בקע קשת תשלום gives equation (iv) as follows:

ערך בקע קשת זוית ב אצל בקע קשת זוית ג הנצבת, הוא כערך בקע קשת תשלום זוית א הנשארת אצל בקע קשת תשלום צלע בג.

(v)	$\text{Cot } A = \text{Cos } c \tan B$	...............(5)	Useful when one other angle, say A, or B, and the hypotenuse C are known.
(vi)	$\text{Sin } b = \text{Sin } c \text{ Sin } B$	................(6)	
(vii)	$\text{Cot } A = \text{Sin } b \text{ Cot } a$	................(7)	

The Talmud contains no mention of spherical triangles or spherical trigonometry, and although Maimonides, in his *Hilchoth Kiddush Hachodesh* (12th century) deals with spherical trigonometry, he does not give any of the properties of spherical triangles. Israeli, however, in his *Yesod Olam* (14th century) gives a full account of spherical trigonometry and deals in detail with the properties of all kinds of spherical triangles.

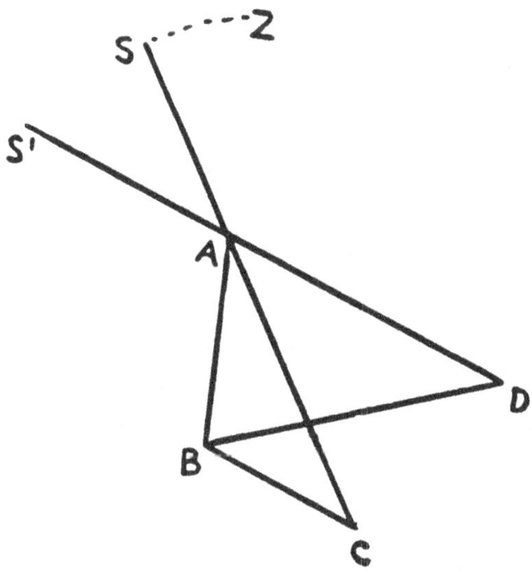

Fig. 33a.

The Gnomon as used by the Ancients for Astronomical Measurments. (Fig. 33) A gnomon is a vertical rod of known height fixed on a smooth horizontal plane, divided into degrees, etc. By measuring the direction and length of the shadow cast by the rod, the ancients were able to make astronomical measurements with considerable accuracy. The following descriptions give the principles, but not the details of the methods:-

[1] Yesod Olam i. 2.

(1) *Altitude of the Sun.* Let AB be the vertical rod, then when the sun is at S, the shadow cast by AB is equal to BC, and the altitude of the sun is the angle BCA. When the sun is at S′ the length of the shadow is equal to BD, and the sun's altitude is the angle BDA. Therefore the higher the sun's altitude the shorter is the shadow, and when the sun is on the Meridian (local noon), i.e., when it reaches the maximum altitude for the day (see p. 68), the shadow is shortest.

(2) *Direction of the Meridian.* In Fig. 33a, BC represents the direction and length of the shortest shadow of the day, and therefore S is on the Meridian. Hence a vertical plane through S and the Zenith Z represents the plane of the local Meridian.[1]

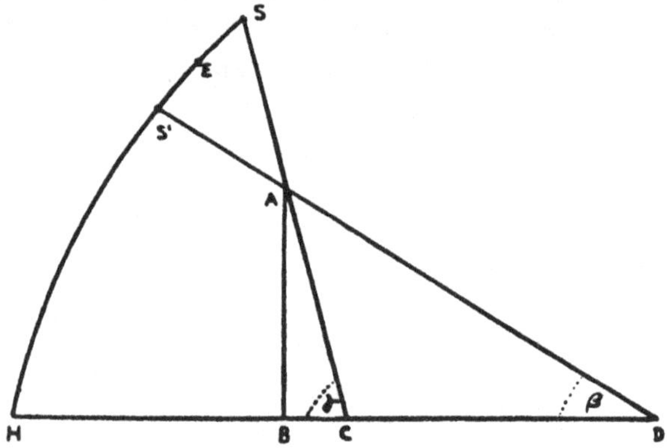

Fig. 33b.

(3) *Declination of the Sun.* Let S (Fig. 33b) be the position of the sun on the Meridian on the day in question, at a place whose geographical latitude is ϕ. Then, if E is the point of intersection of the equator with the Meridian, we have:

Arc HS = altitude of the sun = α,

and Arc HE = colatitude of place = $90 - \phi$

$\therefore$ Declination of Sun, i.e., SE = $\alpha - (90 - \phi)$

(4) *Longitude of the Sun.* Having determined the declination of the sun it is easy by spherical trigonometry to calculate its longitude at that moment. Thus in Fig. 34, if AC = declination of sun = b, and AB = longitude of sun = c, then since angle C is a right angle, and angle B = obliquity of ecliptic = $23\frac{1}{2}°$ approximately, we have from formula vi. p. 83,

$$\text{Sin b} = \text{Sin c Sin B} = \text{Sin c Sin } 23\tfrac{1}{2}° = 0{\cdot}3987 \text{ Sin c}$$

[1] Yesod Olam i. 2.

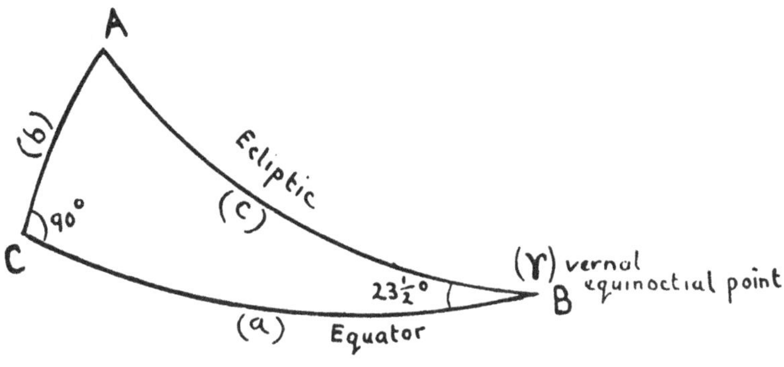

Fig. 34.

$\therefore \quad \text{Sin } c = \frac{\text{Sin } b}{0{\cdot}3987}, \text{ or Sin } \lambda = \frac{\text{Sin } \delta}{0{\cdot}3987}$ (where λ = longitude and δ = declination.)

Hence longitude λ is determined.

(5) *The Moment of an Equinox or a Solstice.* At an equinox, the sun's declination is zero, and at a solstice it has reached its maximum positive or negative value ($\pm 23° \ 27'$). Hence, at noon on a day near the expected event, the sun's declination is determined, and by means of the formula $\text{Sin } \lambda = \frac{\text{Sin } \delta}{0{\cdot}3987}$, the distance of the sun from equinoctial or solstitial point in question, at the time of observation is calculated. As the sun travels at the rate of 0°59′8·33″ a day, or 0°2′28″ an hour (see page 76) the exact moment at which it will reach the equinox or solstice can be found at once.

Example: At noon of March 20th, 1911, the sun's declination was found to be $\delta = 0° \ 17' \ 36''$. What was the exact moment of the Spring equinox תקופת ניסן for that year?

$$\text{Sin } \lambda = \frac{\text{Sin } °0 \ 17' \ 36''}{0{\cdot}3987} = \frac{0{\cdot}0051202}{0{\cdot}3987} = 0{\cdot}0128423$$

$\therefore \quad \lambda = 44'9'' =$ distance of sun from Spring equinoctial point.

This distance will be covered in $\frac{44' \ 9''}{2' \ 28''}$ hours $= \frac{2649''}{148''}$ hours

$= 17$ h. 54 m.

[1] Yesod Olam ii. 4.

Hence the equinox occurred 17h. 54m. after midday of March 20th, i.e., on March 21st at 5.54 a.m.

(6) *Obliquity of the Ecliptic.* With AB=8 units, the ancient Chinese found the following figures (Fig. 33b):

BC i.e., length of Shadow at Summer Solstice (S)= 1·54 units

BD i.e., length of Shadow at Winter Solstice (S′)=13·12 units

$$\therefore \quad \tan ACB = \frac{8}{1\cdot 54} = 5\cdot 1948, \text{ whence arc } SH = 79^\circ\ 6'$$

$$\tan ADB = \frac{8}{13\cdot 12} = 0\cdot 6098, \text{ whence arc } S'H = 31^\circ 22'$$

$$\therefore \quad \text{Arc } SS' = 79^\circ\ 6' - 31^\circ\ 22' = 47^\circ\ 44'$$

But at S the sun is as high above (i.e., north of) the equator as it is at S′ below (or south of) it.

$$\therefore \quad \text{Obliquity} = \frac{47^\circ\ 44'}{2} = 23^\circ\ 52'$$

To find the Altitude of the Pole at any Place. Having found the direction of the Meridian, select a circumpolar star (Fig. 27 on p. 66) and measure its altitude when crossing the Meridian at K and N (i.e., at its upper and lower culminations respectively). The altitude of the pole is seen to be the mean of these two altitudes.

The Altitude of the Pole is equal to the Geographical latitude of the place. At the equator of the earth, the pole is, of course, at the horizon, i.e., when geographical latitude is zero, the altitude of the pole is zero. At the pole of the earth, the celestial pole is, of course, at the zenith, i.e., when geographical latitude is 90°, the altitude of the pole is 90°. Between the equator and the pole, for each degree of geographical latitude the pole rises in altitude by 1°, and therefore at every place on the earth its latitude is at once determined by ascertaining the altitude of the pole. A more general proof can be seen from

Fig 35, where the sphere represents the earth :-

The altitude of the pole (α) is obviously equal to the angle

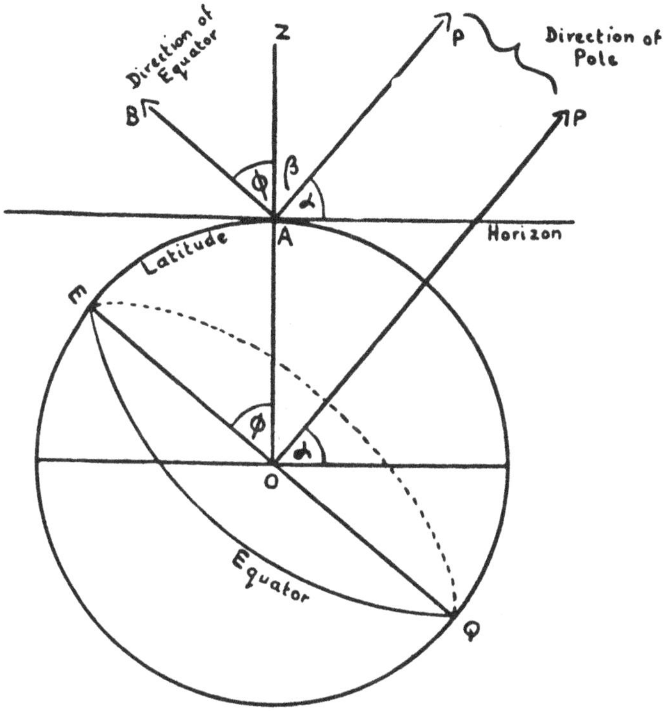

Fig. 35.

ZAB (each being the complement of a right angle). But angle ZAB = AOE = arc AE, which is the latitude ϕ. Maimonides[1], who made his astronomical calculations with Jerusalem as the starting point, gives the latitude of the latter as about 32° North.

כל דרכי החשבון הזה בנוים על עיר ירושלים... ומקום זה הוא נוטה מתחת קו השוה... כנגד רוח צפונית, בכמו שתים ושלשים מעלות.

Israeli[2] gives the latitude of Toledo as 40° North.

Although there is no record of any accurate measurements made by the Talmudic Rabbis by means of the gnomon, there is a very interesting passage[3] which makes it quite evident that they were familiar with the principle of this simple instrument :

הרוצה שלא תשרה חיה רעה בצל קבר, נועץ קנה בד' שעית ביום ויראה להיכן צלו נוטה, משפיע ועולה משפיע ויורד.

[1] Kidd. Hach. xi. 17.

[2] Yesod Olam, ii. 14.

[3] Erubin 43b. & Rashi *ad loc.*

" He who wishes that no animal should seek shelter from the heat of the sun in the shadow cast by a tombstone, should (when he is about to erect it) fix a rod in the ground at 10 o'clock in the morning (when the sun is hottest and the shadow coolest, and is therefore the time of the day when the animals seek the shade) and note the direction of the shadow cast by it. He should then make the stone slant in the plane of that shadow, so as to diminish the length of the shadow."

Note: As a matter of fact, the hottest hour of the day is about 2 p.m., because it is only by that time that equilibrium is obtained between gain and loss of heat by the earth.

The gnomon is utilised in the construction of sundials, which were the time measurers during the day. The latter are mentioned in the Mishnah as אבן שעות[1]. During the night the hours were measured by the ancients by means of the Clepsydra (or Water Clock), or by means of a sandglass. Neither of these is mentioned in the Talmud.

[1] Edyoth iii. 8. & Kelim xii. 4.

CHAPTER IX.

Time of Sunrise and of Sunset. Length of Day.

תנו רבנן לפי שראה אדם הראשון יום שמתמעט והולך, אמר אוי לי שמא בשביל שסרחתי עולם חשך בעדי... כיון שראה תקופת טבת וראה יום שמאריך והולך אמר מנהגו של עולם הוא... כיון שקעה עליו חמה, אמר אוי לי שבשביל שסרחתי עולם חשך בעדי... כיון שעלה עמוד השחר אמר מנהגו של עולם הוא. (עבודה זרה ח.)

The Rabbis taught as follows: When Adam first saw the gradual shortening of the day and the Sun setting in the evening, he said, Woe is me! the world is being destroyed on account of my sin. When he saw that after the Winter Solstice the days began to lengthen again, and that the Sun rose again at dawn on the following morning, he realised that the phenomena were laws of Nature.

Abodah Zarah 8a.

Times of Sunrise and of Sunset. Length of Day. For various religious purposes, e.g., the determination of the beginning and ending of the Sabbath, festival and fast days, as well as for the purpose of ascertaining the true position of the moon at any time (see Ch. xiii.) it is necessary to be able to compute the exact moment of sunset and sunrise at any given place on any given day.

Sunrise (הנץ החמה) is the moment at which the sun appears at the horizon in the morning.

Sunset (שקיעת החמה) is the moment at which the sun disappears at the horizon in the evening.

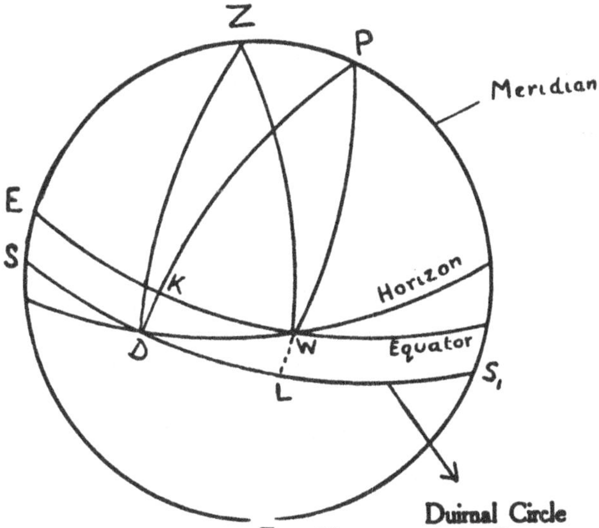

FIG. 36. Duirnal Circle of the Sun.

In Fig. 36, D represents the moment of sunset.

In spherical triangle ZPD, the 3 sides are known:

$ZP = (90 - \phi)^\circ$, where ϕ = geographical latitude of the place.
$ZD = 90^\circ$, and
$PD = PK + KD = PK - DK = (90 - \delta)^\circ$, where δ = declination of sun on the given day.

Therefore, by Corollary to Formula (i), p.82,

$$\text{Cos ZPD} = -\text{Cot}(90 - \phi)\,\text{Cot}(90 - \delta)$$
$$= -\tan\phi\tan\delta.$$

But angle ZPD = Arc SD.

$\therefore$ Cos SD $= -\tan\phi\tan\delta$

$\therefore$ Arc SD is found.

But the sun covers 360° in 24 hours, or 15° in 1 hour, therefore, $\frac{SD}{15}$ gives the number of hours between noon and sunset, and *hence the exact moment of Sunset is known.*

Sunrise occurs $\frac{SD}{15}$ hours before noon, and *therefore its exact moment is known.*

$$\therefore \textit{Length of day} = \frac{2\,SD}{15}$$

Example: Find the time of sunset in Jerusalem (geographical latitude = 32°) on a day when the sun's mean longitude is 60°.

From p.85. we have: $\text{Sin}\,\delta = 0{\cdot}3987\ \text{Sin}\ 60^\circ = 0{\cdot}3987 \times 0{\cdot}8860$
$= 0{\cdot}3453$

$\therefore$ δ (declination of sun) $= 20^\circ 12'$

$\therefore$ Cos SD $= -\tan 32^\circ \tan 20^\circ 12' = -0{\cdot}6249 \times 0{\cdot}3679 = -0{\cdot}2299$

$\therefore$ SD $= 103^\circ, 17', 12'$

and Time of Sunset $= \frac{103^\circ, 17', 12'}{15} = 6$ h. 53 m. 9 s. p.m.

Example. Find the lengths of the longest and shortest days in Jerusalem.

The longest day occurs at the Summer Solstice when $\delta = 23\frac{1}{2}^\circ$.

$\therefore$ Cos SD $= -\tan 32^\circ \tan 23\frac{1}{2}^\circ = -{\cdot}6249 \times {\cdot}4348 = -{\cdot}2717$

Therefore SD $= 105^\circ 46'$, and time of sunset $= \frac{105^\circ 46'}{15}$
$= 7$ h. 3 m. 4 s.

Therefore *length of longest day = 14 h. 6 m. 8 s.*

The shortest day occurs at the Winter Solstice, when $\delta = -23\frac{1}{2}°$

$$\therefore \quad \text{Cos SD} = -\tan 32° \tan (-23\tfrac{1}{2})° = \cdot 2717$$

$$\therefore \quad \text{SD} = 74°14', \text{ and time of sunset} = \frac{74°14'}{15} = 4 \text{ h. } 56 \text{ m. } 56 \text{ s.}$$

Therefore length of shortest day $= 9$ h. 53 m. 52 s.

Hence the difference in the time of Sunset between the longest (or Shortest) day and an equinoctial day, is no more then about 1 *hr. 3 m.*[1]. For the importance of this fact see "**Correction of Moon**".

NOTE—From the formula Cos SD $= -\tan \phi \tan \delta$, we see that the length of the day varies with the geographical position of the place as well as with the declination of the sun on the day in question. But the formula Sin $\delta =$ Sin ϵ Sin λ (p. 85) also tells us that declination depends upon the obliquity of the ecliptic—which as we have seen (p. 69), is gradually diminishing at the rate of about 47″ per century. It follows therefore that the length of the day is gradually diminishing, so that any particular day of the year is somewhat shorter at the present time than the corresponding day, say at the time of the Mishnah—some 1800 years ago—when the obliquity was $18 \times 47'' = 14'$ greater than now. To take an example, we shall compare the length of the day at the Summer Solstice now and 1800 years ago in Jerusalem.

Cos SD, at the present day $= -\tan 32° \tan 23° 27' = -\cdot 6294 \times \cdot 4338 = -\cdot 2730$
whence $2\text{SD} = 211° \cdot 41'$

Cos SD, 1800 years ago $= -\tan 32° \tan 23° 41' = -\cdot 6249 \times \cdot 4386 = -\cdot 2741$
whence $2\text{SD} = 211° \cdot 49'$

Therefore the lengths of the 2 corresponding days are:

$$\frac{211° \cdot 41'}{15} = 14 \text{ h. } 6 \text{ m. } 44 \text{ s., and } \frac{211° \cdot 49'}{15} = 14 \text{ h. } 7 \text{ m. } 16 \text{ s. respectively,}$$

so that that particular day is now about $\frac{1}{2}$ minute shorter than it was then.

Twilight and Dawn. After sunset there is a considerable time, varying in duration with the geographical latitude of the place as well as with the time of the year, before complete darkness sets in. This interval is called *Twilight* (בין השמשות or בין הערבים). The corresponding interval of incomplete darkness before sunrise is called *Dawn* (עלות השחר) These intervals of partial illumination are caused by the reflection of the sun's rays by particles suspended in the air, which thus gives us a certain amount of light when the direct rays cannot yet—or any longer—reach us.

The Talmud[2] discusses whether dawn belongs to the previous night or to the coming day — and comes to no definite decision. Similarly the Rabbis are not quite sure whether twilight belongs to the day or the evening.[3] To be on the safe side, twilight of Friday evening is divided into 3 parts, the first one of which belongs to Friday and the other two belong to the Sabbath day. On Saturday, the first two parts belong to the Sabbath and the other part belongs to the next day (Sunday). Similarly in the case of festival days.

[1] Compare Kidd. Hach. xii. 6.

[2] Beraohoth 8b.

[3] Sabbath 34b. בין השמשות ספק מן היום ומן הלילה.

Duration of Twilight, or Dawn. Astronomically, there are 3 different kinds of twilight, which last different intervals of time, at the same place on the same day. In addition, each kind of twilight lasts a different period at different places or at different times of the year. The *3 different kinds of twilight* are the following:

(a) *Astronomical* (בין השמשות של הכוכבים): From sunset to the moment the centre of the sun has sunk *18°* below the horizon—when minor stars appear (צאת הכוכבים).

(b) *Civil* (בין השמשות של העם). From sunset to the moment the centre of the sun has sunk *6½°* below the horizon—when the zenith loses the red tint (הכסיף התחתון ולא הכסיף העליון). At the end of this twilight an average person is unable to read ordinary print without artificial light.

(c) From sunset to the moment the centre of the sun has sunk *3½°* below the horizon, when the eastern horizon loses its red tint.

It is obvious that each of these kinds of twilight must last a different period of time.

The Talmud gives different estimates for the different kinds of twilight as follows:

איזהו בין השמשות משתשקע החמה, כל זמן שפני מזרח מאדימין והכסיף התחתון ולא הכסיף העליון נמי בין השמשות, הכסיף העליון והשוה לתחתון לילה... שיעור בין השמשות בכמה, אמר רבה שלשה חלקי מיל וכו'...

"What is twilight? The time that the east looks red after sunset. If the zenith has greyed it is still twilight, but if the horizon has also greyed it is night . . . . What is the duration of twilight? Rabah estimates it as the average time it takes to walk ¾ of a Rabbinical mile. R. Joseph estimates it as equal to a walk of ⅔ of a mile, while R. Nehemiah says it is equal to only a ½ mile walk."[1] On the other hand, taking twilight to be the interval between sunset and the appearance of stars משקיעת החמה עד צאת הכוכבים, Rabbah bar bar Channah in the name of R. Yochanan makes it equal to a 5 mile walk, whilst R. Judah estimates it as equal to a 4 mile walk.[2] As the Rabbis estimated that an average person walks 10 parsas (=40 Rabbinical miles) in 12 hours,[3] the different estimates amount to the following times:

[1] Sabbath 34b. & 35a. [2] Pesachim 93b. & 94a. [3] Ibid.

(1) *Rabah*: $\frac{3}{4}$ mile = $13\frac{1}{2}$ minutes		For 3rd kind.
(2) *R. Joseph*: $\frac{2}{3}$ mile = 12 minutes		
(3) *R. Nehemiah*: $\frac{1}{2}$ mile = 9 minutes		
(4) *Rabah bar bar Channah in the name of R. Yochanan*:	5 miles = $1\frac{1}{2}$ hours	For 1st kind.
(5) *R. Judah*:	4 miles = 1 hour 12 minutes	

R. Judah defines twilight as the time between sunset and the appearance of 3 " middle " stars כוכב אחד יום, שנים בין השמשות, שלשה לילה... כוכבים... בינונים.[1]

Causes of Divergences: We shall see that in most cases the divergences can be explained on the following grounds:

(1) Differences in definition of twilight—as seen from the above quotations.

(2) Difference of latitude of the place of observation.

(3) Whether or no the effect of refraction on twilight has been taken into account

Formula for the Duration of Twilight. In Fig. 37, *S*

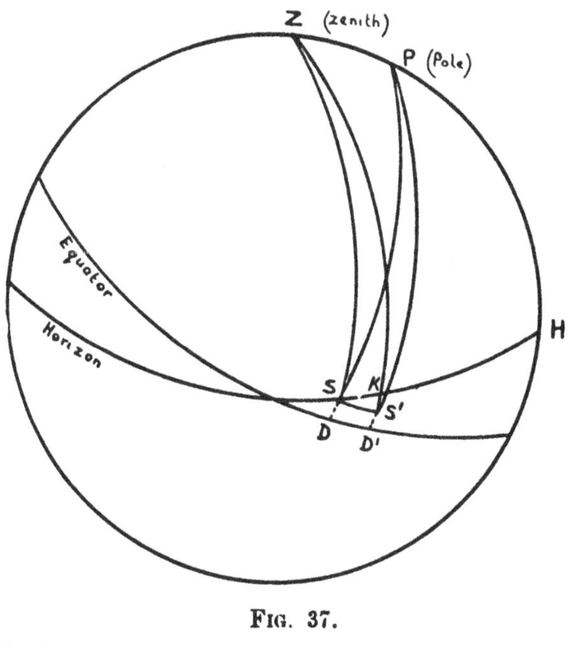

FIG. 37.

[1] Sabbath 35b.

represents the sun at sunset (i.e., on the western horizon), and S′ its position at the end of twilight, i.e., $n°$ vertically below the horizon ($n = 18°$, $6\frac{1}{2}°$, or $3\frac{1}{2}°$, according to the kind of twilight under consideration). We therefore have to find a formula for the time taken by the sun to travel the arc SS′.

In triangle ZPS, the 3 sides are known:

$ZP = (90 - \phi)°$, where ϕ = geographical latitude of place.

$PS = (90 - \delta)°$, where δ = declination of sun on the given day.

and $ZS = 90°$.

Therefore by Corollary to Formula 1, p. 82,

$$\text{Cos } ZPS = -\text{Cot }(90 - \phi)° \text{ Cot }(90 - \delta),° = -\tan\phi \tan\delta$$

Therefore angle ZPS is determined.

Similarly in triangle ZPS′, the 3 sides are known:

$ZP = (90 - \phi)°$

$PS' = PS$ (since SS′ is paralled to the equator $= (90 - \delta)°$

and $ZS' = (ZK + KS') = (90 + n)°$, where n = number of degrees the sun is vertically below the horizon at end of twilight.

Therefore by Formula 1, p. 82,

$$\text{Cos } ZPS' = \frac{\text{Cos }(90+n)}{\text{Sin }(90-\phi)\text{ Sin }(90-\delta)} - \text{Cot }(90-\phi)\text{ Cot }(90-\delta)$$

$$= \frac{-\text{Sin } n}{\text{Cos }\phi \text{ Cos }\delta} - \tan\phi \tan\delta.$$

Therefore angle ZPS′ is determined.

Therefore angle SPS′ (or arc SS′) which $= ZPS' - ZPS$ is known

and Duration of twilight $= \frac{SS'}{15}$.

Taking $\phi = 32$ (Geographical latitude of Jerusalem), and $\delta = 11°45'$ (i.e., the declination of the sun when it is midway between the equinox 0° and the summer solstice $23\frac{1}{2}°$), we get:

(i) $\text{Cos } ZPS = -\tan 32° \tan 11°45' = -·6249 \times ·2080 = -·1300$, whence $ZPS = 97°28'$.

(ii) $$\text{Cos } ZPS' = \frac{-\text{Sin } n}{\text{Cos } 32° \text{Cos} 11°45'} - ·1300 = \frac{-\text{Sin } n}{·8480 \times ·9790} - ·1300$$

$$= \frac{-\text{Sin } n}{·8302} - ·1300$$

Hence the durations of the different kinds of twilight are as shown in the following table.

	$\text{Cos ZPS}' = \frac{-\text{Sin } n}{\cdot 8302} - \cdot 1300$	ZPS′	ZPS	Duration of Twilight $\frac{\text{ZPS}' - \text{ZPS}}{15}$
$n = 18°$	$\frac{-\cdot 3090}{\cdot 8302} - \cdot 1300 = -\cdot 5022$	120° 9′	97° 28′	$\frac{22° 41'}{15}$ = 1 hr. 29 m. 24 s.
$n = 6\frac{1}{2}°$	$\frac{-\cdot 1132}{\cdot 8302} - \cdot 1300 = -\cdot 2662$	105° 27′	97° 28′	$\frac{7° 59'}{15}$ = 31 m. 56 s.
$n = 3\frac{1}{2}°$	$\frac{-\cdot 0610}{\cdot 8302} - \cdot 1300 = -\cdot 2035$	101° 45′	97° 28′	$\frac{4° 17'}{15}$ = 17 m. 8 s.

Effect of Atmospheric Refraction of Light upon the Duration of Twilight. When the rays of light from the sun enter the atmosphere surrounding the earth they are bent in the direction AP shown in the diagram (Fig. 38), so that to an

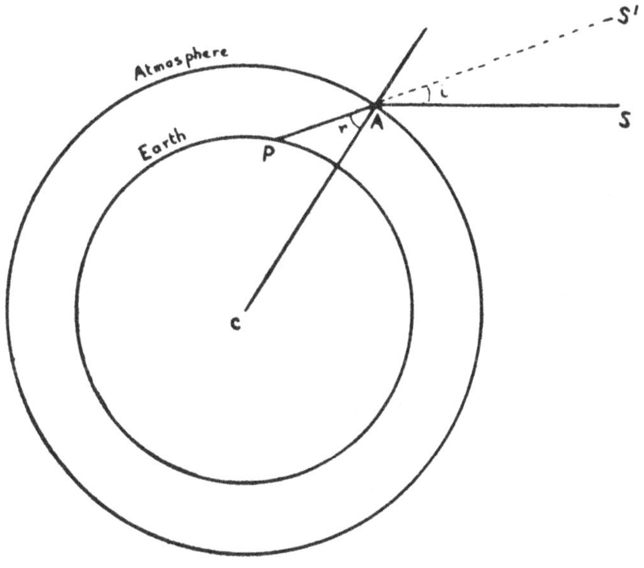

FIG. 38.

observer at P the sun appears to be at S′—which is on the continuation of the line PA—rather than in its true position at S. This bending of the rays of light when entering a medium of different density is called Refraction of Light (עקום האור or שבירת הניצוצות) and its effect is to make celestial bodies appear higher in the sky than they really are. When, therefore, the upper edge of the sun appears at the horizon, it is really some distance below it. Indeed, at the time of apparent sunset, the upper edge of the sun is, as the result of refraction, already 35′ below the horizon. Hence, if we take the sun's semidiameter as 16′, it follows that when the sun just disappears to the observer its centre is already 35′ + 16′ = 51′ below the horizon.

In Fig. 39, S represents the true position of the sun's centre at apparent sunset, viz., 51′ vertically below the horizon, and S′ the position of the centre of the sun at end of twilight, i.e., $n°$

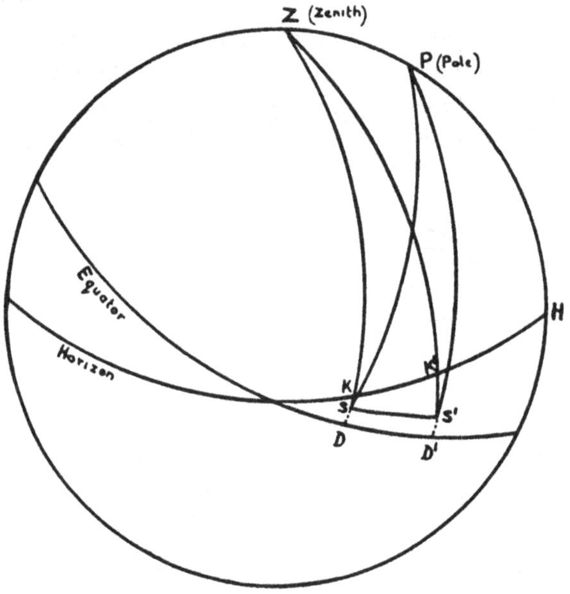

FIG. 39.

below the horizon. Our problem, therefore, is to find the length of the arc SS′.

In triangle ZPS, we have

$$ZP = (90 - \phi)°,\ PS = (90 - \delta)°,\ \text{and}\ ZS = ZK + KS = 90°51'.$$

Therefore from equation 1, p. 82

$$\cos ZPS = \frac{\cos 90°51'}{\sin(90-\phi)\sin(90-\delta)} - \tan\phi\tan\delta$$

$$=\frac{-\text{Sin } 0°51'}{\text{Cos } 32° \text{Cos } \delta} - \tan 32° \tan \delta.$$

Therefore angle ZPS is determined.

Similarly in triangle ZPS′,

$$\text{Cos ZPS}' = \frac{-\text{Sin } n}{\text{Cos } 32° \text{ Cos } \delta} - \tan 32° \tan \delta$$

Therefore angle ZPS′ is determined.

Therefore Duration of twilight, i.e, $\frac{SS'}{15} = \frac{ZPS' - ZPS}{15}$.

Hence, putting $\delta = 11°45'$ as before, we get

(i) $\text{Cos ZPS} = \frac{-\text{Sin } 0°51'}{\cdot 8480 \times \cdot 9790} - \cdot 1300 = -\cdot 1478,$

whence ZPS = 98°30′

(ii) Cos ZPS′ has the same value, for the various values of n, as before. Therefore, taking refraction into account, we get :-

	SS′ = ZPS′ − ZPS	Duration of Twilight
When $n = 18°$	$\frac{SS'}{15} = \frac{120° 9' - 98° 30'}{15} = \frac{21° 39'}{15}$	= 1 hr. 6 m. 36 s.
„ $n = 6\frac{1}{2}°$	$\frac{SS'}{15} = \frac{105° 27' - 98° 30'}{15} = \frac{6° 57'}{15}$	= 27 m. 48 s.
„ $n = 3\frac{1}{2}°$	$\frac{SS'}{15} = \frac{101° 48' - 98° 30'}{15} = \frac{3° 18'}{15}$	= 13 m. 12 s.

Rabah's estimate therefore, of 13½ mins. (which clearly refers to the 3rd kind of twilight) agrees fairly well with mathematical calculation, after allowing for effect of refraction (13m. 12s.).

R. Joseph's estimate of 12 mins, corresponds, to the nearest half-minute, with the calculated duration of the 3rd kind, at the equator at the solstices. Thus

$$\text{Cos ZPS} = \frac{-\text{Sin } 0°51'}{\text{Cos } 23\frac{1}{2}°} = \frac{-\cdot 0148}{\cdot 9170}$$

$$= -\cdot 0161, \text{ giving ZPS} = 0°55'.$$

$$\text{whilst Cos ZPS}' = \frac{-\text{Sin } 3\frac{1}{2}°}{\text{Cos } 23\frac{1}{2}°} = \frac{-\cdot 0610}{\cdot 9170}$$

$$= -\cdot 0665, \text{ giving ZPS}' = 3°49'$$

$$\text{whence } \frac{ZPS' - ZPS}{15} = \frac{3°49' - 0°55'}{15} = \frac{2°54'}{15} = 11\text{m. } 36\text{s.}$$

R. Nehemiah's estimate of 9 mins. is too short for any kind of twilight at any latitude (even at the equator at the equinoxes—when twilight lasts least). For even under such conditions, duration $= \frac{3°30' - 0°51'}{15} = 10\text{m. } 36\text{s.}$

Rabah bar bar Channah's estimate of $1\frac{1}{2}$ hours specifically refers to the 1st kind, and this agrees fairly closely with the calculated 1h. 29m. 24s., when refraction is not taken into account.

Finally, **R. Judah's estimate** of 1 h. 12 m. for the first kind of twilight agrees exactly with the calculated duration of that kind of twilight at the equator (where $\phi = 0$) and at the equinoxes (when $\delta = 0$), without taking into consideration the effect of refraction, since S would then be on the horizon, and S′ would be 18° vertically below it. $\frac{SS'}{15}$ would then be $= \frac{18°}{15} = 1\text{h. } 12$ mins.

Effect of Refraction upon the Length of the Day. Since the effect of refraction is to make the sun appear at the eastern horizon when it is still 35′ below the horizon, and to make it appear just sinking at the western horizon when in fact it is already 35′ below, it is obvious that the apparent length of a day—as measured by the length of time the sun appears to be above the horizon—is longer than the actual day, by the amount of time it takes the sun to cover twice the arc SS′ (in Fig. 37), where KS′ or $n = 35'$. The length of SS′ is found from the formulae on p. 94. in connection with twilight. Thus in the case of Jerusalem ($\phi = 32°$), when $\delta = 11°45'$, we get

$$\text{Cos ZPS} = -{\cdot}1300, \text{ whence ZPS} = 97°28'.$$

$$\text{Cos ZPS}' = \frac{-\text{Sin } 35'}{{\cdot}8302} - {\cdot}1300 = \frac{-{\cdot}0102}{{\cdot}8302} - {\cdot}1300 = -{\cdot}1423$$

whence ZPS′ = 98°11′.

$$\therefore \frac{SS'}{15} = \frac{98°11' - 97°28'}{15} = \frac{0°43'}{15} = 2\text{m. } 52\text{s.}$$

and, therefore, the day is apparently longer by 5 mins. 44 secs.

At the equator, at the equinoxes ($\phi = 0$, and $\delta = 0$), SS′ is vertical and equal to 35′, and the time taken to cover this arc = $\frac{35}{15}$ m $= 2\frac{1}{3}$ mins. Therefore the day, instead of being actually 12 hours long, appears to be 12h. 4m. 40s.

Example. The Day of Atonement in 1930 falls on Thursday, Oct. 2 (i.e., begins on Wednesday evening Oct. 1). Find at what time the fast should begin and end in London (Latitude 51°30′).

Autumn equinox occurs on September 23, when sun's longitude $=180°$.

$\therefore$ On Oct. 1 (i.e., 9 days later),

$$\text{sun's longitude} = 180° + 9 \times 59'8{\cdot}33'' = 188°52'$$

and on Oct. 2 (10 days later),

$$\text{sun's longitude} = 188°52' + 59'8{\cdot}33'' = 189°51'$$

Therefore the declinations δ_1 and δ_2 of the sun on these days are given by

$$\text{Sin}\,\delta_1 = 0{\cdot}3987 \text{ Sin } 188°52' = -0{\cdot}3987 \text{ Sin } 8°52'$$
$$= -0{\cdot}3987 \times 0{\cdot}1541 = -0{\cdot}05144$$
$$\text{Whence } \delta_1 = -2°57'$$

Similarly, $\text{Sin}\,\delta_2 = -0{\cdot}3987 \text{ Sin } 9°51' = -0{\cdot}3987 \times 0{\cdot}1711$
$= -0{\cdot}0682$, whence $\delta_2 = -3°55'$

$\therefore$ Cos ZPS (Fig. 39) on Oct. 1 $= -\tan 51°30' \tan(-2°57')$
$= \tan 51°30' \tan 2°57' = 1{\cdot}2572 \times 0{\cdot}0515 = 0{\cdot}0647$
Whence ZPS $= 86°17'$

and Cos ZPS on Oct. 2 $= \tan 51°30' \tan 3°55' = 1{\cdot}2572 \times 0{\cdot}0685$
$= 0{\cdot}0861$, whence ZPS $= 85°4'$

$\therefore$ *Time of Sunset or beginning of twilight on Oct. 1* $= \dfrac{86°.\ 17'}{15}$
$= 5$ h. 45 m. 8 s.

and Time of Sunset on Oct. 2 $= \dfrac{85°.\ 4'}{15} = 5$ h. 40 m. 16 s.

We next find duration of twilight on each of these days.

End of twilight on Oct. 1 is given by,

$$\text{Cos ZPS}' = \frac{-\text{Sin } 18°}{\text{Cos } 51°30' \text{ Cos } (-2°57')} + {\cdot}0647$$
$$= \frac{-{\cdot}3090}{{\cdot}6225 \times {\cdot}9987} + {\cdot}0647 = -{\cdot}4326$$

$\therefore$ ZPS$' = 115°38'$

$\therefore$ *End of twilight* is at $\dfrac{115°38'}{15} = 7$ h. 42 m. 32 s.

$\therefore$ Duration of twilight on Oct. 1 = 7 h. 42 m. 32 s. – 5 h. 45 m. 8 s.
= 1 h. 57 m. 24 s.

$\therefore$ *Fast begins* at 5 h. 45 m. 8 s. + $\frac{1}{3}$(1 h. 57 m. 24 s.)
i.e., at *6h. 24 m. 16 s.*

End of twilight on Oct 2 is given by

$$\text{Cos ZPS}' = \frac{-\text{Sin } 18^\circ}{\text{Cos } 51^\circ 30' \text{ Cos } 3^\circ 55'} + \cdot 0861$$

$$\text{Cos ZPS}' = \frac{-\text{Sin } 18^\circ}{\text{Cos } 51^\circ 30' \text{ Cos } 3^\circ 55'} + \cdot 0861$$

$$= \frac{-\cdot 3090}{\cdot 6225 \times \cdot 9977} + \cdot 0861 = -\cdot 4124$$

$\therefore$ ZPS′ = 114°21′

$\therefore$ *End of twilight* $= \dfrac{114^\circ 21'}{15}$ = *7 h. 37 m. 24 s.*

$\therefore$ Duration of twilight = 7 h 37 m. 24 s.– –5 h. 40 m. 16 s.
= 1 h. 57 m. 8 s.

Fast ends at 5 h. 40 m. 16 s. + $\frac{2}{3}$(1h. 57 m. 8 s.)
i.e., at *6 h. 59 m. 19 s.*

Jewish Time. The hour in Jewish literature is of two kinds, viz.:

(1) *The Astronomical or Equinoctial hour,* שעת השווי, which is $\frac{1}{24}$ of a complete day (i.e., of a day and night מעת לעת).

(2) *The Civilian hour,* שעת זמנית which is $\frac{1}{12}$ of a complete period of daylight.

Whilst the Astronomical hour is always of equal length, the civilian hour varies from day to day and becomes equal to the Astronomical hour at each equinox.

The day begins with sunset. At dawn the 12th night-hour has finished and the 1st day-hour begins. Hence, when the Mishnah says, for instance, that the Tamid sacrifice is to be offered at 8½ hours[1] "תמיד נשחט בשמונה ומחצה וכו that time is not a fixed one, but varies with the season and from day to day. At the equinoxes, when sunrise is at 6 a.m., astronomical time, it would correspond to 2.30 p.m. On the other hand, when sunrise is at 4 a.m. astronomical time (and sunset is at 8 p.m.) then daylight lasts 16 hours, and one civilian hour = $\frac{16}{12}$ astronomical hours = 1 hour 20 mins. The time 8½ hours would therefore mean $8\frac{1}{2} \times 1\frac{1}{3}$ astronomical hours = 11hr. 20m. after 4 a.m. = 3.20 p.m.

Division of the Hour. The Rabbis divided the hour not sexagesimally into 60 minutes consisting of 60 seconds each, but into 1080 parts or Chelakim (חלקים) each of which was divided

[1] Pesachim v. 1.

into 76 moments or Regaim (רגעים). One minute, therefore, corresponds to $\frac{1080}{60} = 18$ Chelakim or to $18 \times 76 = 1{,}368$ regaim. Thus R. Gamliel gives the time of a complete lunation (i.e., the interval between two moons, or what is commonly called a synodic month) as $29\frac{1}{2}$ days, $\frac{2}{3}$ hour and 73 Chelakim.[1] As 18 Chelakim $= 1$ minute, therefore 73 Chelakim $= 4\frac{1}{18}$ minutes $=$ 4m. $3\frac{1}{3}$s. Therefore the synodic month is 29d. 12h. 44m. $3\frac{1}{3}$sec.—the value given by Hipparchus and Ptolemy.[2] (See *Mean Conjunction*, Chapter xii.)

Origin of this form of Division of the Hour. Maimonides[3] explains this peculiar method of dividing the hour on the basis that 1080 is a multiple of 2, 3, 4, 5, 6, 8, 9, 10 לפי שמנין זה יש בו חצי, ורביע, ושמינית, ושליש, וששית, ותשע, וחומש, ועישור. but inasmuch as the number 360 would equally satisfy these requirements, and as, moreover, the number does not include a multiple of 7, the explanation is unsatisfactory. Mahler's explanation is a very reasonable one[4]: The Babylonians estimated the value of the precession as equal to 1° in 72 years, so that a precessional year—i.e., the time of a complete revolution of the equinoctial points $= 72 \times 360 = 25{,}920$ years. They therefore divided the ordinary day of 24 hours into 25,920 parts, and each hour consequently consists of $\frac{25920}{24} = 1080$ parts or Chelakim. For the reason of dividing the Chelek into 76 regaim, see under "**Metonic Cycle**".

[1] Rosh Hashonah 25a.
[2] See Yesod Olam iii. 2.
[3] Kidd. Hach. vi 2.
[4] Handb. d. jüd. Chronol., Leipzig, 1916, p. 27.

CHAPTER X.

The Earth.

תולה ארץ על בלימה (איוב כ"ו ו").

He hangeth the Earth upon nothing. (Job xxvi. 7).

In the earlier part of the Talmudic period, i.e., till about the end of the 2nd century C.E., the Jews, in common with others, believed the earth to be flat, and to be suppported by pillars—the exact number of which was disputed. Thus R. Josi of the 2nd century, basing himself on Job iv. 6.: " Which shaketh the earth out of her place and the pillars thereof tremble " (ועמודיה יתפלצון) infers that the earth stands on pillars. Certain anonymous sages, on the basis of Deut. xxxii. 8, fixed the number at twelve. Others, basing themselves on Proverbs ix. 1, reduced it to seven, whilst R. Elazar b. Shamua (a contemporary of R. Josi), probably realising that the position of the earth in space could not be explained on such a theory, quotes Proverbs x. 25 (וצדיק יסוד עולם) in support of his statement that the earth rests upon one " pillar ", viz., righteousness. The pillars were made to rest on Water, the Water on Mountains, the Mountains on the Wind, the Wind on the Whirlwind and the Whirlwind was supported on the arm of God.[1]

ר' יוסי אומר... הארץ על מה עומדת על המים... מים על ההרים... הרים ברוח... רוח בסערה... סערה תלויה בזרועו של הקדוש ברוך הוא... וחכמים אומרים על י"ב עמודים... ויש אומרים ז' עמודים... ר' אלעזר בן שמוע אומר על עמוד אחד וצדיק שמו וכו'.

It was not till the middle of the 4th century that R. Jonah stated that the earth was spherical. ראה העולם כבדור.[2] We know, of course, that it is supported by nothing.

The Size of the Earth.

(1) *Circumference.* Raba, of the 4th century gives the length of the earth's circumference as being 6,000 parsah.

אמר רבא שיתא אלפי פרסי הוי עלמא.

A parsa, or parsang, is a Persian mile and is equal to about 4·1454 English miles, and therefore Raba's estimate amounts to 24,872 miles, which is practically identical with that of modern estimates.

[1] Chagigah 12b [2] Jer. Ab. Zarah iii.

Raba's figure is, however, almost certainly taken from Ptolemy, who found the length of 1° of terrestrial latitude to be 500 stadia. As a stadium is $\frac{1}{30}$ parsa, therefore 1° terrestrial latitude $= \frac{1}{30}$ parsas, and hence, the whole circumference $= \frac{50}{3} \times 360 = 6,000$ parsas, which is identical with the figure given by Raba.

Indeed, it is possible that even Ptolemy may have taken this figure from Eratosthenes, who in the 3rd century B.C. estimated the earth's circumference with remarkable accuracy. Aristotle, a century earlier still, gave a figure which is 60 per cent too high. Israeli[1] gives the length of half the equatorial circumference as 12,000 miles, which is equal to a 400 days journey. This, he rightly continues, makes the radius equal to 3,818 miles, which latter he says is taken as the unit or " measuring rod " of astronomical measurement (קנה המדה)

(2) *Radius.* As the circumference $= 2\pi r$, where $r =$ radius and $\pi = \frac{22}{7}$ therefore $2\pi r = 6,000$ parsas and $r = \frac{6,000 \times 7}{44} = 954 \cdot 5$ parsas $= 3,964$ miles.

(3) *Surface Area of the Earth.* The area of a sphere $= 4\pi r^2$; therefore area of earth is $4 \times \frac{22}{7} \times (954 \cdot 5)^2 = 14,310,597$ square parsas. This is about $\frac{1}{4}$ of the estimate given by the Talmudic Rabbis, who make the area of the world " 60 times that of Aethiopia, which latter is 60 times that of Egypt, whose area is 160,000 square miles."

תנו רבנן, ארץ מצרים הוא ארבע מאות פרסה על ארבע מאות פרסה, והוא אחד מששים בכוש, וכוש אחד מששים בעולם.

This makes the area of the earth $160,000 \times 60 \times 60 = 576,000,000$ square parsas.

The way the circumference of the earth (and hence its radius and area) was estimated by the ancients, was to measure the altitude of the pole at two places, on the same meridian, whose distance from each other is known. As the difference in altitude of the pole measures the difference in latitude of the two places (see p. 87), the length of 1° of latitude is ascertained. This multiplied by 360 gives the length of the circumference.

Such measurements taken at different parts of the earth's surface have shown that a degree of latitude is not of constant length, being least near the equator (68.7 miles) and most, near the pole (69·4 miles), which shows that the earth is not quite

[1] Yesod Olam ii. 3.

spherical but flattened at the poles—making the polar diameter about 26 miles shorter than the equitorial. This, however, was not known to the ancients.

Distribution of Land and Water. During the Middle ages before the Western Hemisphere was discovered, the Jews, in common with others, believed one half of the globe to be covered by water (the Great Ocean ים אוקינוס) and the other half to be dry and inhabitable. Thus R. Abraham bar Chiya of the 11th century begins his work on the Jewish Calendar (ספר העבור) by stating that "the earth is round like a sphere. The water of the Great Ocean covers half of it, which is uninhabitable, the other half is dry and inhabitable." He goes on to say that the total area of the islands in the midst of the ocean is equal to that of the lakes in the midst of the dry land, with the result that exactly half the area of the globe is dry land and half is water; in proof whereof he quotes Gen. I., 9 and 10, and continues by saying that God divided every item in the creation into two equal parts. Thus, in the same way as there is as much daylight as night in the year, so half the earth is dry land and the other half is water, "and there is no need to produce any other proof of such an obvious fact"! The Tosefoth (Pesachim 94) divides the globe into three equal parts, viz., seas, deserts and habitable land.

Isaac Israeli of the 14th century in *Yesod Olam* (II. 1) expresses similar views to those of R. Abraham bar Chiya, based on similar evidence and disagrees with the view that there is a fiery sphere surrounding the atmosphere — on the sole ground that there is no mention made of it in the story of the creation as narrated in Genesis!

The zones of longtitude (רצועות הישוב) were called *Climates* (אקלימים) -of which there were seven, the seventh being the northern limit of the inhabited eastern hemisphere. "Jerusalem and the whole of Africa are in the 4th climate; Egypt is in the 3rd climate; Toledo and several Spanish provinces are in the 5th climate," etc. (Yesod Olam ii. 3).

CHAPTER XI.

The Sun's Apparent Annual Motion.

הרואה חמה בתקופתה, לבנה בגבורתה וכוכבים במסלותם ומזלות כסדרם, אומר ברוך עושה מעשה בראשית. (ברכות נ"ט:)

"He who looks at the Sun in his circuit, the Moon in her power, the Stars in their courses, and the planets in their systems, must pronounce: "Blessed be He who has created the universe."

Berachoth 59b.

Although the curve traced out by a *projection* of the sun's path on the celestial sphere is, as we have seen, a great circle—called the ecliptic—it does not follow that the sun's *actual* orbit is a circle of which the earth is the centre. Whatever the shape of that path—whether circular or elliptical, a regular or irregular curve–provided it is always in the same plane, it will be projected on the surface of the celestial sphere as a great circle, since every plane of any shape passing through the centre of a sphere cuts the surface of the sphere in a great circle. As a matter of fact it was known to Hipparchus in the 2nd century B.C. and even, possibly, before his time, that the true path of the sun's orbit could not be a circle of which the earth is the centre, because instead of its rate of movement, as seen by an observer on the earth, being perfectly constant—as should be the case on that hypothesis[1]—the sun is observed, by actual measurement of its daily increase in longitude, (see p. 84) to move more rapidly during some portions of the year than at other times. Also, on the same supposition, the apparent diameter of the sun should be constant throughout the year—its distance from the earth being constant — but modern refined measurements have shown, what Hipparchus with his more primitive apparatus could not detect, that it is larger at some periods of the year than at others — the maximum difference amounting to about 1′. As the apparent diameter is inversely proportional to the distance, it follows that the sun must be at different distances from the earth at different times of the year, its maximum distance being when its apparent diameter is least, and its minimum when the diameter is greatest. We

[1] See Kidd. Hach. xi. 13.

מהלך שנה אין בו לא קלות ולא כבדות אלא כמו מהלכו היום כמו מהלכו אמש כמו מהלכו למחר וכו'.

know now, since Kepler's time, that the sun's true orbit is an ellipse (of eccentricity 0·0168) of which the earth is one of the foci, but to the ancients the simplest explanation that presented itself was that the orbit was really circular, but that the earth was eccentrically situated within it, i.e., the centre of the orbit did not coincide with that of the earth.[1] Such a circle is called an eccentric (גלגל יוצא המרכז)[2]

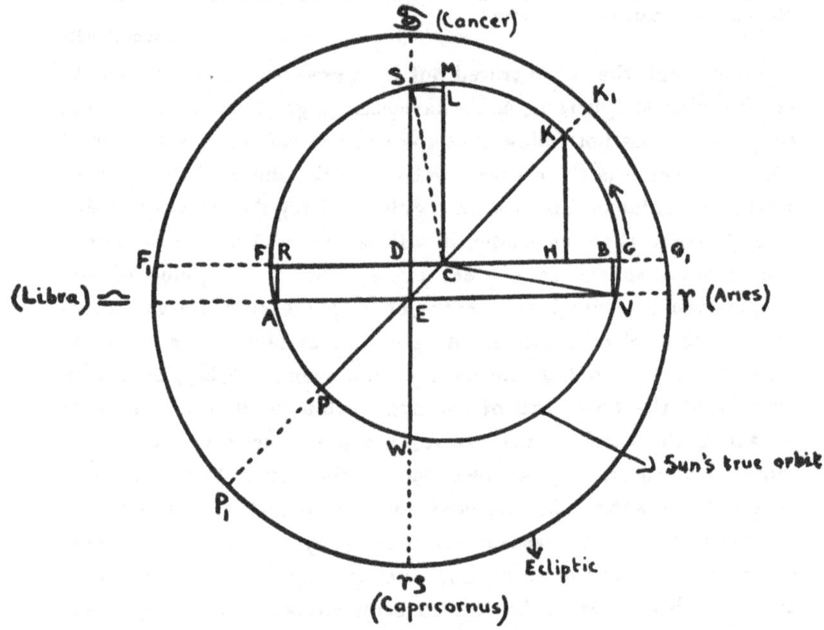

FIG. 40.

Thus in Fig. 40 the internal circle, of which C is the centre, represents the sun's actual orbit, in which the centre of the earth is eccentrically situated at E. The external circle, of which E is the centre, is the ecliptic—or the projection of the sun's actual orbit on the celestial sphere. At the vernal equinox the sun is actually at V in its true orbit but at ♈ in the ecliptic; S and ♋ represent the sun's position at the Summer Solstice in the true orbit and on the ecliptic respectively; A and ♎ the respective positions at the autumnal equinox; and W, ♑ the two positions at the Winter Solstice. The line PECK, passing through the centre of the earth (E) and that of the sun's orbit (C) is the *Line of Apsides* (קו המישור) or *Apse Line*[3] (Apsis = joining or fastening).

[1] Kidd. Hach. xi. 13. הגלגל אף ע"פ שהוא מקיף את העולם אין הארץ באמצעו.
[2] Israeli, Yesod Olam ii. I. [3] Ibid III. 4. See also Ibid i 4

Now, whilst the sun actually describes the big arc VGKMS in its true orbit (i.e., round centre C), it appears to an observer on the earth (at E) to describe the quadrant (i.e., the arc of 90°) from ♈ to ♋ on the ecliptic. Similarly, whilst it actually describes the small arc (i.e., an arc less than 90°) of SA in its true orbit (i.e., round centre C), it appears to describe the quadrant (90°) ♋ ♎ in the ecliptic (round centre E). Assuming the sun's velocity to be uniform the time taken by the sun to travel from V to S must therefore be greater than that taken to travel from ♋ to ♎ so that to an observer at E, the sun appears to cover 90° of the ecliptic in a longer time in the spring (Aries to Cancer) than in the summer (Cancer to Libra), thus seeming to move more slowly in the spring than in the summer. Similarly, it appears to move faster in the autumn than in the winter. Indeed, comparison of the arcs VS and AW will show that the sun appears to travel in the ecliptic slowest in the spring and fastest in the autumn. The sun also appears to travel slower between the beginning of spring and the beginning of autumn, i.e., between ♈ and ♎ than between the beginning of autumn and that of spring, i.e., between ♎ and ♈. Similar inequality of speed occurs between any other two diametrically opposed points of the ecliptic, except between the points K_1 and P_1, i.e., between *Apogee* (רום החמה[1] גובה השמש[2]) when the sun is furthest from the earth E and *Perigee* (שפל החמה)[1] when it is nearest the earth, because the apse line P_1PECKK_1, passing as it does through the centres of the true orbit and of the ecliptic, bisects each of these circles, so that the arcs K_1♋P_1, P_1♑K_1, KSP and PWK are all equal to one another = 180°[3].

Eccentricity of the Sun's Orbit. (יציאת המרכז, מנת המרכזים) or (ערך הקו שבין שני המרכזים)[4] By the eccentricity of the orbit (e) is meant the ratio $\frac{EC}{CK}$ i.e., the ratio of the distance between the two centres to the radius of the orbit. Its value, a knowledge of which is necessary for our purpose, can be easily calculated as follows:

[1] Israeli l. c. iii. 1.

[2] Maimonides l. c. xii. 3.

[3] As we know now that it is the earth which moves round the sun and not the sun round the earth, Apogee is now called *Aphelion* (away from the sun) and Perigee is called *Perihelion* (near the sun).

[4] Israeli l. c. iii 6.

In Fig. 40, triangles CED and CKH are similar.

$$\therefore \frac{EC}{CK} = \frac{ED}{KH} = \frac{VB}{KH} = \frac{VB/CV}{KH/CK} = (\text{since } CV = CK) = \frac{\text{Sin VG}}{\text{Sin KCH}} = \frac{\text{Sin VG}}{\text{Sin KEV}}$$

We shall show in the next paragraphs how the values of VG and angle KEV can be calculated. Assuming that in the time of Maimonides VG was 1°58′, and KEV was 86°45′8″, we get:

$$\text{Eccentricity } \frac{EC}{CK} = \frac{\text{Sin } 86°45'8''}{\text{Sin } 1°58'} = \frac{0{\cdot}0343}{0{\cdot}9984} = 0{\cdot}0345 \text{ (i.e., about } \tfrac{1}{30}\text{)}$$

Once the eccentricity of the orbit is determined, the position of the centre C of the orbit is fully determined if we also know the angle CEV or KEV (i.e., longitude of apogee), which we now proceed to ascertain.

Evaluation of Arc VG and of Angle KEV. (Fig 40)

(1) *Arc VG.* Al Battani in 880 C.E. found (by measurement of the sun's declination, see p. 84.) spring and summer to last 93d. 14h. and 93d. $\frac{3}{4}$h. respectively, so that at the rate of $\frac{360°}{365\frac{1}{4}}$ or 0°59′8·3″ per day the sun covers during those seasons:

$$\text{Arc VGS (Spring)} = 93\tfrac{7}{12} \times 0°59'8{\cdot}3'' = 92°14'21{\cdot}740''$$

$$\text{Arc SFA (Summer)} = 93\tfrac{1}{32} \times 0°59'8{\cdot}3'' = 91°41'42{\cdot}785''$$

$$\therefore \text{ Whole Arc VSA} = 183°56'\ 4{\cdot}525''$$

$$\text{Subtract Arc GSF} = 180°\ 0'\ 0{\cdot}000''$$

$$\therefore \text{ VG+FA, i.e., 2VG} = 3°56'\ 4{\cdot}525''$$

$$\text{Whence VG} = 1°58'\ 2{\cdot}260''$$

(ii) *Angle KEV.* Since KEV = DCE (Fig. 40)

$$\therefore \textit{Tan KEV} = \tan \text{DCE} = \frac{DE}{DC} = \frac{BV}{SL} = \frac{BV/CV}{SL/CS} (\text{Since CV} = \text{CS})$$

$$= \frac{\text{Sin BCV}}{\text{Sin SCL}} = \frac{\text{Sin GV}}{\text{Sin SM}}$$

As GV is known, KEV can be computed if we also know SM. But SM = VGS − (GM + VG) = 92°14′21·74″ − 91°58′2·26″
= 0°16′19·48″

$$\therefore \tan \text{KEV} = \frac{\text{Sin VG}}{\text{Sin SM}} = \frac{\text{Sin } 1°58'2{\cdot}26''}{\text{Sin } 0°16'19{\cdot}48''} = \frac{{\cdot}03433}{{\cdot}00466} = 7{\cdot}36695$$

$$\therefore \text{KEV} = 82°16'11''$$

Al Battani gives it as 82°17′ (quoted by Ibn Junus as 82°14′). Ibn Junus from personal observation made KEV = 86°10′.

The positions of Apogee and Perigee, and therefore also of the Apse Line, are determined by observing the positions of the sun in the heavens, and the exact time of those positions when the sun has its smallest and largest apparent diameters. (See p. 105.)

Motion of the Apse Line. (תנועת נקודת הגובה) Maimonides gives the value of the longitude of the sun's apogee (i.e., the angle KEV) at the Epoch (22nd March, 1178) as 86°45′8″ or 26°45′8″ in sign Gemini[1] ומקום גובה השמש היה בעיקר זה בשש ועשרים מעלות חמשה וארבעים חלקים ושמנה שניות ממזל תאומים.

He also states that this longitude increases at the rate of about 1° in 70 years, or, more precisely, at the rate of 1·5″ in 10 days (i.e., 1° in 65·7 years),[2] which is approximately the rate of precession ומהלכה בכל שבעים שנה בקירוב מעלה אחת... בכל עשרה ימים שניה וחצי.

These two statements raise two interesting points. Firstly, assuming the increase in Apogee's longitude to be 1·5″ in 10 days, then in view of the fact that Al Battani on 1st March, 880, made this angle equal to 82°17′, its value on 22nd March, 1178—i.e., after an interval of 108865 days — should have been 82°17′+ $\frac{108865 \times 1 \cdot 5''}{10}$ = 82°17′+4°32′10″=86°49′10″ and not 86°45′8″. It is, therefore, difficult to see how Maimonides arrived at the value he gives—unless he did not rely on Al Battani's figure and estimated the longitude of the sun's apogee from actual personal observation.

Secondly, we now know that in addition to the increase of the longitude resulting from precession, i.e., at the rate of 50·25″ per year, the angle KEV increases also at the rate of about 11·25″ per year as the result of the *Independent motion of the Apse line* in the same direction as the sun (i.e., opposite to that of precession) —making the total annual increase of KEV equal to 61·5″. As this independent motion of the Apse line was recognised by Zarkali (1080) before the time of Maimonides, it is somewhat strange that the latter should have ignored this component. Indeed, a comparison of the longtitude of Apogee as found by Ptolemy (KEV= 65°32′), and 750 years later by Al Battani (viz. 82°17′) shows that longtitude of apogee had in the interval increased by

[1] Op. Cit. xii. 3. [2] Ibid xii. 2.

$\frac{82^\circ 17' - 65^\circ 32'}{750}$ = about 80″ per annum, i.e., by about 30″ more than could be accounted for by precession. Rabbi Isaac ben Sid in 1240, i.e., 360 years after Al Battani gives the position of the sun's apogee as 90°—i.e., at the beginning of the sign Cancer, or at the summer solstice. This is the position adopted by Israeli[1] who gives the Apse line an independent motion of 1° in 280 years or about 13″ per year from west to east, and is the position in which apogee should have been in that year on the supposition that its increase in longitude is at the rate of 80″ per year. Thus $82^\circ 17' + 80'' \times 360 = 90^\circ 17'$.

NOTE: Owing to the gradual increase in the angle KEV, i.e., of longitude of apogee, the duration of the seasons keeps on changing. Thus:

In the time of Ptolemy	Durations were:	Spring $94\frac{1}{2}$ days,	Summer $92\frac{1}{4}$ day.
.. .. Al Battani	.. .. ..	$93\frac{7}{12}$ ..	.. $93\frac{1}{12}$ ..
In 1910	.. .. ..	$92\frac{5}{8}$ ..	.. $93\frac{3}{8}$ ..

The last values make VG (Fig. 40) = $1^\circ 52' 40{\cdot}2''$ and SM = $-0^\circ 22' 40{\cdot}7''$, whence KEV = $101^\circ 22'$, and eccentricity = ·0334 = $\frac{1}{30}$ If we take Al Battani's date (880 C.E.) as the Epoch, then at the rate of increase of 61·5″ per year, the value of KEV should be $82^\circ 17' + 1030 \times 61{\cdot}5'' = 99^\circ 51' 40''$— which is nearly the same as $101^\circ 22'$.

NOTE ON THE SUN'S ECCENTRIC ORBIT. Although a circular orbit of eccentricity = $\frac{1}{30}$ explains the irregular speed of the ecliptic, such a curve cannot be the true shape of the orbit. For were that so, we should find the apparent diameters of the sun at perigee P and apogee K and in the proportion EK : EP = $\frac{31}{29}$ = 1.07 : 1. Neither Hipparchus nor Ptolemy could measure such small differences in the sun's diameter, but modern refined measurements have shown that the longest and shortest apparent diameters of the sun are 32′.36″ (Dec. 31) and 31′.32″ (July 1) i.e. in the proportion of 1·034 : 1, from which it follows that perigee and apogee occur on those two respective dates, and that $\frac{1+e}{1-e} = 1{\cdot}034$, (where c = eccentricity) and therefore e = 0·0167 i.e. about $\frac{1}{60}$.) Hence a *circular eccentric orbit cannot explain the sun's path.* As a matter of fact, the orbit is an *ellipse* of eccentricity 0·0168, and the sun's, or rather the earth's, *actual rate of motion in the orbit is not uniform,* being slower near apogee (aphelion) than near perigee (perihelion). Its slowest and quickest rates of motion have been found to be 0°.57′.11·5″ (July 1) and 1°.1′.9·9″ (Dec. 31) per day respectively i.e. in the proportion of 1 : 1·069.

Hence during the time that the apparent diameter has changed in the ratio of 1·034 : 1 the angular velocity has changed in the ratio of 1 : 1·069 or 1 : $(1{\cdot}034)^2$ i.e. the sun's angular velocity is inversely proportional to the square of its distance from the earth. (See Kelper's Laws, Chapter xiv.)

[1] Yesod Olam iii. 2.

Position of the Sun in the Heavens at any Time. To know the exact position of the sun on the celestial sphere at any moment is a matter of very great importance in Rabbinical Astronomy, because such knowledge enables one to find by easy—if rather complicated—mathematics the exact position of the moon at that moment. The importance of knowing the latter lies in the fact that during the Talmudic period the beginning of a new month was determined on the accredited evidence of witnesses, who testified to having seen the new moon, and in order to test the reliability of such witnesses it was necessary for the Calendar Council to ascertain by mathematical calculation whether at the time the witnesses alleged to have seen the moon, the latter was in such a position in the heavens as to be visible. The nature of such calculations is never mentioned in the Talmud, but Maimonides in his *Hilchoth Kiddush Hachodesh* (הלכות קדוש החדש) published in 1178 gives a large number of working rules, as well as an abundance of numerical data, by means of which such calculations can be accomplished, but he gives no explanation of the principles on which these rules are based, or of the manner in which the numerical data have been obtained. Israeli (Rabbi Isaac ben Joseph Israeli of Toledo, a pupil of R. Asher ben Yechiel —הראש) in his Yesod Olam (ספר יסוד עולם), written in 1310, gives a very complete illustrated account of Ptolemaic Astronomy, and a diligent mastery of that work helps one considerably to grapple with the very difficult chapters of Maimonides' very important, but most difficult, book. In addition, it is impossible to speak too highly of Prof. E. Baneth's monumental monograph on "*Maimunis' Neumondsberechnung*", which, by giving the mathematical principles upon which Maimonides bases his calculations, presents the interested reader with the key which opens what must otherwise remain a Pandora's box to the vast majority of scholars. To Baneth's classical work as well as to Hechim's, שבילי דרקיע[1] I wish to acknowledge my very great indebtedness.

Data required for calculating the Sun's and Moon's position in the Heavens. (יסודי המסלול)

A. The Sun's Position.

(1) Sun's longitude at a given starting date—the so-called Epoch or Origin (שנת חעיקר)

[1] Szwile Derokia, edited by B. S. Lewenstein, Warsaw, 1863.

(2) **Sun's period of rotation, i.e., the length of the Tropical year (called by Israeli (שנה תקופית) or the interval between two successive vernal equinoxes. This gives the arc covered by the sun in one day or portion of a day.**

(3) **The Eccentricity of the Sun's true orbit, i.e., the value of the ratio $\frac{EC}{CK}$ in Fig. 40 (ערך שבין שני המרכזים or מדת יציאת המרכזים or מנת המרכזים)**

(4) **The longitude of the Sun's Apogee, i.e., the longitude of the sun when it is furthest away from the earth (angle KEV in Fig. 40).**

B. The Moon's position.

Once these data are known, the sun's position in the ecliptic becomes easily ascertainable (See p. 114), *and when the latter has been found, the moon's position on the celestial sphere can be computed if we know in addition :*

(5) The inclination of the Moon's orbit to the ecliptic.

(6) The longitude of the Moon's ascending node (Ch. xii).

(7) The Moon's period of rotation.

Finally, having thus ascertained by calculation the moon's position in the heavens at any given moment, (Chapter xiii.) it is necessary to determine whether in that position, it would be visible by an observer on the earth ; and for determining this Visibility the following two additional data are necessary, viz.

(8) The Elongation—or difference in longitude between Sun and Moon.

(9) The Arc of Visibility—or difference in altitude between these two bodies. (See Chapters xiv. and xv.).

1. The Epoch (העיקר שממנו מתחילין) adopted by Maimonides[1] is sunset of Wednesday, 22nd March, 1178 (i.e., the beginning of Thursday, 3rd Nisan, 4938, world era). The year is that of the publication of his *Hilchoth Kiddush Hachodesh*, and the day of the month was chosen because it was very near the vernal equinox, and it also corresponded to the earliest time after a conjunction (new moon) when the moon's crescent became visible.

The sun's longitude at that moment is given by Maimonides[2] as having been 7°3′32″, and the longitude of the sun's apogee (i.e., size of angle KEV) at the same moment was 86°45′8″.[2]

[1] op. cit. xi, 16. [2] Ibid xii. 3.

2. Mean Rate of Sun's Motion. (מהלך השמש האמצעי) is given by Maimonides as roughly 0°59′8″ per day — or more exactly as 98°33′53″ in 100 days, which corresponds to 0°59′8·33″ per day,[1] 2′27·85″ per hour or 2·46″ per minute. (The Nautical Almanac for 1929 gives the mean rate of the sun's motion as 0°59′8·34″ per day.)

This makes the *tropical year* $= \dfrac{360°}{0°59'8{\cdot}33''} =$ 365d. 5h. 48m. 49·59s. This value is correct within $\frac{1}{5}$ second, as can be seen from the following calculation:

Sidereal year (שנה תמימה) i.e., time taken by the sun to return to the same position relative to the fixed stars = 365·2563582 days (= 365d. 6h. 9m. 9·35 s.) Now the value of precession in the year 1178 was 50·1000528″, therefore the tropical or equinoctial year at that time (i.e., the interval between two successive vèrnal equinoxes) must have been less than the sidereal year by the time the sun would take to cover an arc of 50·1000528″. But the sun covers an arc of 1° in $\dfrac{360}{365{\cdot}2563582}$ days.

∴ It would cover 50·1000528″ in $\dfrac{365{\cdot}2563582}{360 \times 60 \times 60} \times 50{\cdot}1000528$ days = 20·3326 mins. = 20m. 19·96 secs.

Hence, the tropical year in the time of Maimonides was 365d. 6h. 9m. 9·35s.– 20m. 19·96s. = 365d. 5h. 48m. 49·39s., which is only $\frac{1}{5}$ second less than that given by Maimonides. It is remarkable that 365 years later Copernicus made the tropical year 365d. 5h. 49m.12s., which is 20 seconds higher than its true value.

Ibn Ezra[2] states that the ancient Hindus found the length of a sidereal year to be 365d. 6h. 12m.

לא פירש משה מהלך שנה תמימה, כי חכמי המזלות לא יכלו עד הנה להוציאה לאור, כי חכמי הודו מוסיפים על רביע היום חומש שעה.

Calculation of the Position of the Sun on the Ecliptic at any moment. Knowing the daily rate of motion of the sun in its true orbit (which, of course, $= \dfrac{360°}{\text{number of days in year}}$) we can easily find its position in that orbit at any moment, if we know what its position in the orbit was at any given moment, called

[1] Kidd. Hach. xii. 1. It is to be noted that this is not the true mean rate of the sun's motion, but its mean rate of separation from the vernal equinox. Its own independent motion is only at the rate of 0°59′8′1[illegible]″ a day.

[2] Commentary to Levit. xxv. 9.

the *Epoch*. From the point of view of Rabbinical Astronomy, however, the important point to know is not the sun's position in its true orbit, but where it would appear (to an observer on the earth) to be on the ecliptic, or celestial sphere. Fig. 41 will

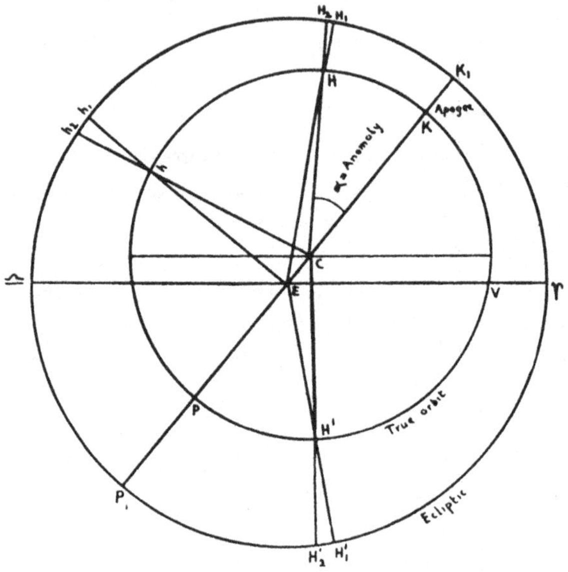

FIG. 41.

explain how the sun's position on the ecliptic at any moment can be calculated from a knowledge of its position in the true orbit.

When the sun is at H in its true orbit, it would appear to an observer at the centre of the orbit, C, to be on the ecliptic at the point H_2, which is the point where the continuation of the line CH cuts the ecliptic,[1] but to an observer on the earth, E, it appears to be at H_1 where the line EH, continued, cuts the ecliptic.[2] The point H_1 is called the sun's *true position* or *true longitude* (מקום השמש האמתי) and H_2 is the suns' *mean position* (מקום השמש האמצעי) We have seen that the sun's mean position moves uniformly at the rate of 59′8·33″ a day, but that its true position changes irregularly. If, however, we look at the diagram in Fig. 41 we see that the sun's true position H_1 can be found if we know its mean position H_2 simply by subtracting from, or

[1] The line CH is called the Radius Vector קו המניע (Israeli op. cit. iii. 6.)
[2] The line E is called the Indicator קו המורה (Ibid)

adding to, its position, the value of the arc H_1H_2, according as the sun is on one or other side of the apse line.

If the *Angular distance of the sun from the apogee,* i.e., angle HCK, *called the Anomaly*-or מסלול החמה is less than 180°, then its true longitude γH_1 is less than its mean longitude γH_2—so that the arc H_1H_2 has to be subtracted from the mean longitude, to find the true longitude. If, however, the anomaly is greater than 180°, e.g. when the sun is at H′ in its true orbit, so that arc H′PK>180, then the sun's true longitude $\gamma K_1H'_1$ is greater than its mean longitude $\gamma K_1H'_2$, and the arc $H'_2H'_1$ has to be added to the mean longitude to find the true longitude.

It therefore remains merely to find the value of the arc H_1H_2 which is the same as the angle H_1HH_2 or EHC. This angle varies with the anomaly, i.e., with the angle HCK, and can easily be computed if the anomaly at any moment is known. For in the triangle EHC, EC and CH are known—or rather the ratio between them—is known (=eccentricity (e)=0·0347 : 1)—and angle ECH is known (=180°- anomaly), therefore all other parts of the triangle are known, so that angle EHC, or Arc H_1H_2 is determined.

From Plane trigonometry we know that

$$\frac{CH-EC}{CH+EC}\text{ i.e. }\frac{1-e}{1+e}=\frac{\tan\frac{1}{2}(HEC-EHC)}{\tan\frac{1}{2}(HEC+EHC)}$$

$$=\frac{\tan\frac{1}{2}(HEC-EHC)}{\tan\frac{1}{2}\text{ Anomaly}}\quad(\text{since }HEC+EHC=HCK=\text{Anomaly})$$

$$\therefore\quad\frac{1-{\cdot}0347}{1{\cdot}0347}\text{ or }0{\cdot}9329=\frac{\tan\frac{1}{2}(HEC-EHC)}{\tan\frac{1}{2}\text{ Anomaly}}$$

$$\text{or }\tan\tfrac{1}{2}(HEC-EHC)=0{\cdot}9329\tan\frac{\alpha}{2}\ (\text{Where }\alpha=\text{Anomaly})$$

Hence, if anomaly is known, HEC - EHC is at once found.

We then have $HEC-EHC=\beta$ (say)

and $HEC+EHC=\alpha$

$\therefore$ By subtraction $2EHC=\alpha-\beta$

$$\therefore\quad EHC=\frac{\alpha-\beta}{2}$$

The anomaly is of course calculable at any moment, since it is equal to the sun's mean longitude at that moment, less the longitude of its apogee at the same moment—each of which can be ascertained at any moment in the manner we have described (pp. 84 and 109).

The angle EHC is called the **"Equation of the Centre"**; Maimonides[1] calls it the quota of the anomaly (מנת המסלול) since it varies with the anomaly. Israeli[2] calls the arc H_1H_2 "the Variation Arc" (קשת החילוף), and the angle EHC, which is equal to it "the Variation Angle" (זוית החילוף) and we therefore say that the **Sun's true longitude = Sun's mean longitude + Equation of centre.**

Example: Find the value of EHC, when the anomaly is 30°.

$$\text{Tan } \tfrac{1}{2}(\text{HEC} - \text{EHC}) = 0{\cdot}9329 \tan 15° = 0{\cdot}9329 \times 0{\cdot}2679$$
$$= 0{\cdot}2499$$
$$\therefore \tfrac{1}{2}(\text{HEC} - \text{EHC}) = 14°2'$$
$$\therefore \text{HEC} - \text{EHC} = 28°4'$$
$$\text{But HEC} + \text{EHC} = 30°$$
$$\therefore \text{by subtraction, } 2\text{EHC} = 1°56', \text{ whence EHC} = 0°58'0''$$

This value agrees exactly with that given by Maimonides (see Table on p. 117). The angle H_2CK_1, or arc H_2K_1, is called the mean anomaly, and the angle H_1CK_1, or arc H_1EK_1, the true anomaly. Therefore, **True Anomaly = Mean Anomaly ± Equation of centre.**

The values given by Maimonides agree very well with those calculated on the supposition that the orbit is represented by an eccentric circle, of eccentricity 0·0347. The orbit is, as we now know, an *ellipse* of eccentricity 0·01675, but the values of the equations of the centre for different values of the anomaly are practically identical with those given by Maimonides. The following table on p. 117 compares the values of EHC (equation of centre) as given by Maimonides[3] (a), as calculated on the theory of an eccentric circular orbit (b), and as calculated on the theory of an elliptical orbit (c). The Formula for the equation of the centre in an elliptical orbit is approximately 114·6° e Sin a, where e = eccentricity and a = mean anomaly. In the case of the sun, therefore, where e = 0·01675, the expression becomes 1·92° Sin a.

It will be seen from the table that the **maximum equation of the centre is** when the anomaly hEK is 90°—viz., **1°59'.** This is also seen from the diagram Fig. 41, where the maximum equation of the centre is the angle EhC, whose Sine is $\frac{\text{EC}}{\text{Ch}} = 0{\cdot}0347$. Such an angle is 1°59'.

[1] Kidd. Hach. xiii. 2. [2] Yesod Ol.m. iii. 1. [3] Kidd. Hach. xiii. 4.

Anomaly (α) (מסלול)	Equation of Centre (מנת המסלול)		
	(a) Maimonides	(b) Calculated for eccentric circular orbit (as on p 116)	(c) Calculated for elliptical orbit from formula 1·92 Sin α
10°	0°20′	0°20′ 0″	0°20′ 0″
20°	0°40′	0°39′28″	0°39′24″
30°	0°58′	0°58′ 0″	0°57′36″
60°	1°41′	1°41′28″	1°40′34″
90°	1°59′	1°59′16″	1°55′20″
120°	1°45′	1°44′58″	1°40′34″
150°	1°1′	1° 1′25″	0°57′36″
170°	0°21′	0°21′25″	0°20′ 0″

Intermediate values are found by the method of *Proportional Parts*[1] as in the example below.

If the anomaly is greater than 180°, then you take its complement. Thus if anomaly = 300°, then you subtract 300° from 360°, leaving 60°, and its quota, or equation of the centre, is 1°41′. As, however, the anomaly is greater than 180°, this quota must be added to, and not subtracted from, the mean longitude, to obtain the true longitude.

The method of finding the sun's true position at any moment from a knowledge of its mean position at that moment is called **Correction** or Adjustment **of the sun** (תיקון החמה)

Example : Maimonides works out the method of finding the sun's true longitude at the beginning of Saturday, 14th Tammuz, 4938 W.E., corresponding to sunset of Friday, 29th June, 1178, as follows :[2]

At Epoch, sunset of Wednesday, 22nd March, 1178, the sun's mean position was 7°3′32″ and the longitude of its apogee was 86°45′8″.[3] Therefore on 29th June, 1178, i.e., 100 days later, its longitude increased by $100 \times 0°59'8{\cdot}33''$, i.e., by 98°33′53″, to 105°37′25″, and the longitude of apogee increased by 15″ to 86°45′23″. Hence the mean anomaly on the date in question = $105°37'25'' - 86°45'23'' = 19°$, approximately.

But equation of centre for an anomaly of 10° = 20′, and that for an anomaly of 20° is 40′. Therefore equation of centre for anomaly of $19° = 20' + \frac{(40' - 20') \times 9}{10} = 38'$.

1 Kid. Hach. xiii. 7. 2 Ibid. xiii. 8 & 9. 3 Ibid. xiii. 3.

As the anomaly is less than 180°, therefore true longitude = mean longitude − equation of centre

$$= 105°37'25'' - 0°38'0''$$
$$= 104°59'25''$$
$$= 105°, \text{ approximately.}$$

Therefore the sun was, on the date in question, 15° in the sign of Cancer.

If we take modern data, viz., increase of longitude of apogee at the rate of 17″ in 100 days, and the equation of the centre on the basis of an elliptical orbit, we get:

Longitude of Apogee on the date in question $= 86°45'8'' + 0°0'17''$
$= 86°45'25''$

∴ Mean Anomaly $= 105°37'25'' - 86°45'25'' = 18°52'$

Equation of the centre, for an elliptical orbit, is $1{\cdot}92 \text{ Sin } 18°52' = 37'15''$. Hence true longitude $= 105°37'25'' - 37'15'' = 105°0'10''$, which is practically identical with the 104°59′25″ given by Maimonides.

This mean position, as Maimonides rightly points out[1], refers to the place of the sun an exact number of complete days after the moment of the epoch. But as the sunsets of the two days in question do not occur at exactly the same moment of the day, it is strictly speaking necessary to adjust for this difference. The difference, however, in the case of Jerusalem, is so small as to be negligible, for as we saw on p. 90, the difference between the times of sunset of the longest (or shortest) day and an equinoctial one, is no more than about 1 hour, during which solar longitude increases, by no more than about 2½′ which is inappreciable. It is, continues Maimonides, when we come to calculate the position of the moon, whose motion is more than 13 times as fast as that of the sun, and where, therefore, a difference of an hour would alter the moon's longitude by about $2\frac{1}{2}' \times 13 =$ about 33′, that differences in the times of sunset on the two days in question must be taken into account.

האמצעי שיש בחשבון זה פעמים יהיה בתחלת הלילה בשוה או קודם שקיעת החמה בשעה או אחר שקיעת החמה בשעה ודבר זה לא תחוש לו בשמש שהרי אנו משלימים קירוב זה כשנחשב לאמצע הירח.

Different kinds of Astronomical Year. An astronomical year is the period of complete revolution of the sun (or earth) from a given point in the heavens to the same point again. There are, therefore, as we have seen, several kinds of year varying in length with the starting point, according as the latter is stationary or moving; and if moving, whether its motion is in the same or in the opposite direction, as follows:-

[1] Kidd. Hach. xii. 6.

Kind of Year	Starting Point		Duration
	name	rate and direction of motion	
Sidereal (שנה תמימה)	Fixed Star	0	365·256382d. = 365d. 6h. 9m. 9·35s.
Tropical (שנה תקופית)	Vernal Equinox	50·1″ a day *to meet* sun	365d. 5h. 48m. $49\frac{1}{3}$s.
Anomalistic	Apogee	11·25″ a day *same direction* as sun	365d. 6h. 13m. 48s.
Ecliptic	Node (point of intersection of plane of ecliptic with that of lunar orbit)	3′11″ a day *to meet* sun	346·62 days = 346d. 14h. 52m. 48s.

A Civil Year contains an exact number of days as follows:-

Solar: contains 365 days and in leap year 366 days.

Lunar: contains 353, 354, or 355 days (common years) or 353, 354, or 355 days (leap years). See p. 187.

CHAPTER XII.

The Moon.

כתיב עשה ירח למועדים שמש ידע מבואו (תהלים ק"ד, י"ט), שמש הוא דידע מבואו ירח לא ידע מבואו. (ראש השנה כ"ה).

"It is written: "He appointed the Moon for the Seasons, the Sun knoweth his going down" (Ps. civ. 19). This implies that the Sun knows his course, but the Moon knows not hers." (Rosh Hashanah 25a). —

The Lunar Orbit. The moon's path, as projected on the celestial sphere is a great circle, but—as in the case of the sun (p. 105) — its irregular velocities in different parts of its course (ranging from 27′ to 33′40″ per hour), as well as the variations in its apparent diameter from 29′30″ to 33′40″ prove conclusively that its orbit cannot be a circle of which the earth is the centre. Indeed, as in the case of the sun and all the planets, the moon's orbit is now known to be an ellipse (see Ch. XIV.), although Ptolemy explained its orbit's very marked irregularities in a much more complicated manner—as we shall see later (p. 132).

The Nodes (נקודות התלי)[1] Measurement of the moon's latitude

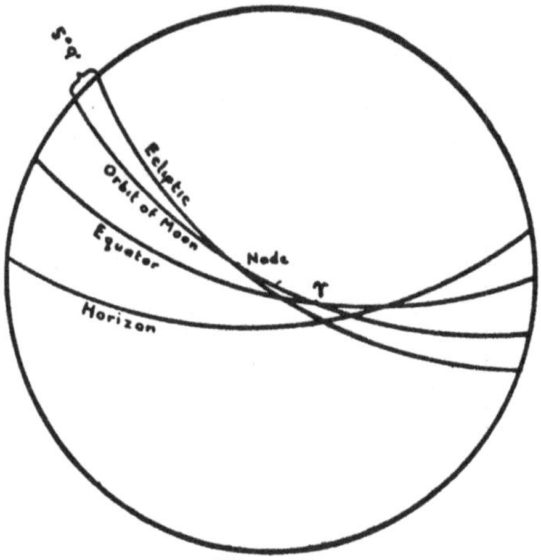

Fig. 42.

[1] Yesod Olam, iii. 8.

(i.e., its angular distance from the ecliptic) in different parts of its course, show that its orbit is inclined to the ecliptic at an angle whose size is known to vary between 4°57′ and 5°20′—with a mean value of about 5°9′ (Fig. 42). Maimonides[1] and Israeli[2] give its value as 5°—as found by Ptolemy. The points of intersection of the planes of the lunar and solar orbits are called the *Nodes*, and the straight line joining them is called the *line of the Nodes*. It is obvious that since an eclipse of the sun or moon can only occur when these two bodies lie in the same straight line as the earth, such a phenomenon can only happen when the moon

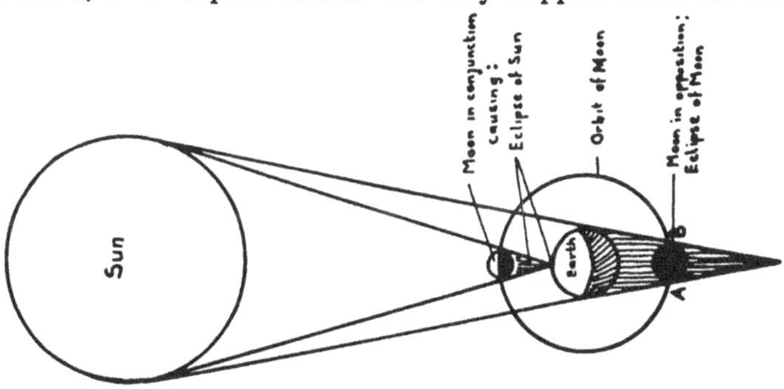

Fig. 43

is at, or near, one or other node (Fig. 44). Now the ancient astrologers believed that the cause of an eclipse was a dragon *(draco)*, which stretched across between the nodes, and which devoured the sun or moon when either of these crossed its path—a new luminary being born after every such catastrophe. They, therefore, called the nodes the *head* and *tail of the dragon* (ראש וזנב התנין)[3] and the line of the nodes was the dragon itself. The node at which the moon passes from south to north of the ecliptic is

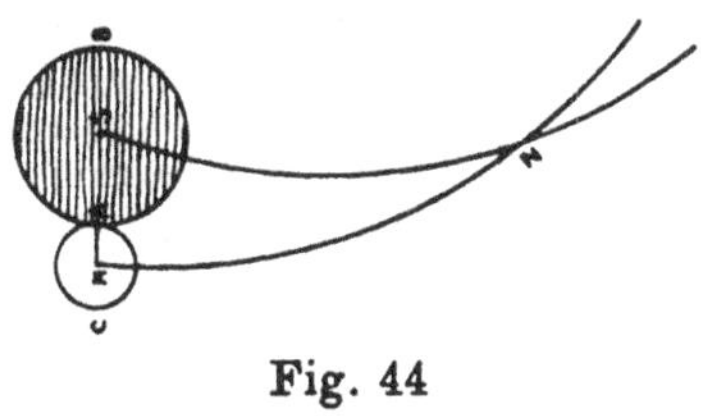

Fig. 44

called the *Ascending Node* (head of the dragon), and the corresponding node on the downward path from north to south is the *descending node* (tail of the dragon).

Kidd. Hach. xvi. 9. [2] Yesod Olam, iii. 8.
[3] Yesod Olam, iii, 8. Maimonides calls them simply the head and the tail. (Kidd. Hach. xvi. 1)

Revolution of the Nodes (גלגול התלי אחורנית).[1] The nodes, like the equinoctial points, have a retrograde (East to West) motion at a rate which Maimonides gives roughly as 3′11″ a day, but more exactly as 5°17′43″ in 100 days—i.e., at the rate of 3′10·63″ a day,[2] which is about 1400 times as fast as the rate of precession, viz., 0·14″ a day (p. 77), and the *period of a complete revolution of the nodes* is therefore only about $\frac{1}{1400}$ of a precessional year, i.e., $\frac{26000}{1400} = 18.6$ *years.*

The rate of movement of the nodes has been measured by ascertaining the longitude of the moon each time it crosses the ecliptic—i.e. each time its latitude becomes zero. As the sun's daily rate of movement in the opposite direction to that of the node is 59′, 8·33″ (p. 113), the sun and node separate at a daily rate of about 62′.19″. They will therefore meet again in $\frac{360^\circ}{62'.19''} = 346.62$ days. This period is called the *synodic period* of the nodes, or an *Ecliptic Year.*

Elongation of the Moon. (מרחק הירח), The moon makes a complete revolution of the heavens (360°) in relation to the equinox, in 27d. 7h. 43m. 4s.—a period known as a *Tropical Month.* Its average daily rate of motion from West to East is, therefore, 13°10′35·03″,[3] corresponding to 32′56·5″ per hour, or about 33″ per minute. As the sun's average daily rate of motion in the same direction is 59′8·33″ the separation between these two heavenly bodies is 13°10′35·03″ − 59′8·33″ = 12°11′26·70″ per day. The angular distance (measured along the ecliptic), or difference in longitude between the moon and sun at any moment is called the *Elongation* of the Moon, so that Elongation increases at the rate of 12°11′26·70″ a day, or 30′28·6″ per hour or 30·5″ per minute.

Conjunction (קבוץ). When elongation is zero, the moon is said to be in conjunction (New Moon), and it is then between the sun and the earth. In Rabbinical Astronomy there are two kinds of conjunction to be considered :-

(i) *True Conjunction* (קבוץ האמתי), i.e., the exact moment at which the centres of the sun and moon actually have precisely the same longitude. As neither the sun's nor the moon's rate of motion is uniform, the interval between the two consecutive

[1] Yesod Olam, iii, 2. [2] Kidd. Hach. xvi. 2. [3] Kidd. Hach. xiv. 1 & 2.

conjunctions, i.e., the length of a *lunation*, or a *Synodic Month*[1] is not constant, but varies within a few hours round about a mean interval of 29d. 12h. 44m. 3⅓s. At those times of the year when the sun's true motion is rapid, the elongation increase is less, and therefore the month is longer than the mean synodic period. When the sun's motion is slow, the month is shorter.

The Talmudic Rabbis recognised the variation in length of the synodic month[2], אמר ר׳ גמליאל כך מקובלני מבית אבי אבא פעמים שבא בארוכה פעמים שבא בקצרה, and hence they determined the beginning of every month separately by observation of the new moon as well as by calculation. This is called the *True Conjunction* — or *Phase—Method* of Calendar Making.

ii. *Mean Conjunction* (קבוץ האמצעי), or *Molad* (מולד), i.e., the moment that the moon would have the same longitude as the sun, if both moved uniformly. Since the time of the Talmud—i.e., during the last 1500 years—the Phase—or Observation—Method of Calendar Making קביעות ראש חדש על פי הראיה[3] was abandoned, and the *Fixed Calendar Method* קביעות ראש חדש על העיבור[4] was adopted—a method which does not concern itself with the true, but with the mean, conjunction, or Molad. As the increase in mean daily elongation is 12°11′26·7″, it follows that the length of a mean synodic month is $\frac{360^\circ}{12^\circ 11' 26{\cdot}7''}$ = 29d. 12h. 44m. 3s., so that if the time of a mean conjunction of any given month is known, the Molad for any subsequent (or previous) month can easily be found (see under **"Fixed Calendar"**)

R. Gamliel is credited with having given the correct duration of a mean synodic month[5] אמר להם ר׳ גמליאל כך מקובלני מבית אבי אבא אין חדושה של לבנה פחותה מעשרים ותשעה יום ומחצה ושני שלישי שעה וע״ג חלקים.

" R. Gamliel said to them : I have it on the authority of my father's father that a synodic month cannot be less than twenty nine days and half a day, two thirds of an hour and 73 chelakim " (=29d.12h.40m. + $\frac{73}{1080}$ × 60 min. = 29d.12h. 44m. 3⅓s).

In view of the fact that this figure is the one given by Ptolemy, it would not be surprising if R. Gamliel was acquainted with it. Nevertheless it seems practically certain that the words "and 73 chelakim", in the above quotation

[1] Synod = a meeting (between sun & moon.)

[2] Rosh Hashanah 25a. [3] Kidd. Hach. v. i. & 3. The Talmud uses the expression קיבועא דירחא for the Phase method (see Succah 43b.)

[4] Ibid. vi. i. [5] Rosh Hashanah 25a.

are a comparatively recent interpolation, for as David Gans[1] points out — not only would R. Gamliel have given 793 chelakim (חשצ"ג חלקים), instead of "$\frac{2}{3}$h. 73ch.", but in the Boraitha of Samuel as quoted by R. Abraham bar Chiya, as well as the Pirkè d.'R. Eliezer, (which R. Abraham bar Chiya believes to be identical with the Boraitha of Samuel), the words "and 73 chelakim" are missing. The Boraitha says that the lunar (synodic) month contains 7384$\frac{2}{3}$ hours (=29d. 12$\frac{2}{3}$h). (See Sefer Haibbur, ii. 2.)

כל שעות של חרש הלבנה תש"ח שעות ושתי ירות שעה.

Further, Gans points out, that it is hardly likely that R, Gamliel would speak of a *mean* synodic month, as in his time the "Fixed Calendar" method was not yet in use.

Slonimsky[2] goes as far as to say that even the words "and a half day and $\frac{2}{3}$ hour" are interpolations. Indeed, were R. Gamliel to speak of the *mean* length of a synodic month, there is no point in his saying that it cannot be less than..."; he should have said that "its duration is.... ". It is only when one speaks of the *true* length of the month — which, as we have seen, is variable, — that one can give a minimum value,

Israeli[3] gives the maximum interval between a true and a mean conjunction (of the same month) as 15 hours. אינו אפשר ולא יכול שיקדם רגע המולד על עת הקיבוץ האמיתי שלעוכתו או שיתאחר ממנו אלא בפחות מט"ו שעות.

Opposition (הקבלה) is the moment when the elongation is 180°. The moon is then full, and the moment of opposition is exactly the middle of the month. The earth is then between the sun and the moon. In both conjunction and opposition the moon is said to be in *syzygy*.

Quadrature (רבוע) is the phase of the moon when its elongation is 90° or 270°. It is half moon and occurs at the end of the first or third quarter of the month.

Eclipses. Were it not for the inclination of the plane of the moon's orbit with that of the ecliptic (p. 121), there would be an eclipse of the sun (לקוי חמה) at every conjunction, or new moon —because of the opaque moon coming between the sun and the earth and cutting off the sun's light; and an eclipse of the moon (לקוי לבנה) at every opposition, or full moon—because of the earth coming between the sun and the moon, and casting a shadow upon the latter (Fig. 43). Owing, however, to this inclination, no eclipse can occur unless at the time of conjunction or opposition the moon happens to be at, or near, one or other node, i.e., when the two orbital planes actually or nearly intersect.

[1] נחמד ונעים, Yessnitz 1743, §.213, (pp. 63b. and 64a)
[2] יסודי העבור, 3rd. ed., Warsaw 1883 p. 34. [3] Yesod Olam, iii, 11.

Ecliptic Limits (גבולי הליקוי)[1]. In Fig. 44, the large black circle AB represents the shadow of the earth, whose centre is on the ecliptic, and AB is its diameter (corresponding to AB in Fig. 43). Hence AS is the shadow's semi-diameter. Also, the small circle CA represents the moon, and MA is the moon's semi-diameter. Hence a lunar eclipse is bound to occur so long as the moon's latitude (i.e., its distance from the ecliptic), is not greater than MS (i.e., than the sum of the semi-diameters of the moon and of the earth's shadow). When the moon's latitude exceeds that sum, no eclipse can take place. The longest distance NS, of the moon from the node, measured along the ecliptic, at which a lunar eclipse can occur, is called the *Lunar Ecliptic Limit.*

Major & Minor Ecliptic Limits. Since the distance of the sun from the earth varies in different parts of its course—therefore the arc AB of the earth's shadow, and hence also its semi-diameter AS, must vary from a maximum at aphelion (apogee) to a minimum at perihelion (perigee). Allowing for parallax and refraction, the semi-diameter of the earth's shadow has a maximum of 46′45·3″ and a minimum of 37′0·25″. Similarly the moon's apparent semi-diameter varies between a maximum of 16′50″ at perigee and a minimum of 14′48″ at apogee. We thus have :-

(1) Maximum Value of MS (when sun is at aphelion and moon at perigee) = 46′45·3″ + 16′50″ = 1°3′35·3″ which means that if the moon's latitude is greater than this, no eclipse is possible.

(2) Minimum Value of MS (sun at perihelion and moon at apogee) = 37′0·25″ + 14′45″ = 51′45·25″ which means that if the moon's latitude is less than this, an eclipse is inevitable.

The value of NS for the maximum latitude of the moon compatible with an eclipse is called the *Lunar Major Ecliptic Limit,* and that for the minimum latitude of the moon, below which an eclipse is inevitable, is called the *Lunar Minor Ecliptic Limit.*

Calculation of the Ecliptic Limits. As the arc SM and the angle at N of the spherical triangle SNM are very small, the triangle may for practical purposes be treated as a plane one, and the angle at M may be taken as a right

[1] Yesod Olam, iii, 17.

angle. Although angle SNM varies between 5′. 20′ & 4′. 57, thus producing an extra variation in the ecliptic limit, we may with sufficient accuracy take SNM as equal to 5°.

We thus have $\frac{SM}{NS}$ = Sin SNM = Sin 5° = 0 0·872, ∴ NS = $\frac{SM}{0{\cdot}0872}$

∴ When SM = 1°3′3537, we get NS = $\frac{63{\cdot}388'}{0{\cdot}0872}$ = 12° 9′ = *Major Limit*

and when SM = 51′18·25″, we get NS = $\frac{51{\cdot}304''}{0{\cdot}0872}$ = 9°44″ = *Minor Limit.*

Israeli[1] gives the values of the Major and Minor limits as 12°12′ and 11° respectively.

To find whether in any given month a lunar eclipse will take place. We must first find the exact moment of a true opposition in that month (see next chapter). The next step is to ascertain the true latitude of the moon at that moment. (see next chapter.)

Having found the latitude, we can at once find the distance of the moon from the node—as measured along the ecliptic. If that distance is less than the minor ecliptic limit i.e. less then about 10°, an eclipse is inevitable. If it is greater than the major limit of about 12$\frac{1}{6}$° an eclipse is impossible. If it is between these two values, then it is necessary to find the apparent semi-diameter of the moon, as well as the semi-diameter of the earth's shadow at that moment and compare their sum with the moon's latitude. If the latter is greater than this sum, no eclipse is possible; if it is less, an eclipse is inevitable.

Solar Ecliptic Limits. The maximum and minimum values of NS consistent with a solar eclipse are called the major and minor solar ecliptic limits.

They are 18°30′ and 15°20′. respectively.

Time Duration and Character of a Solar Ecliptic.

It is possible to compute all the details about a solar eclipse as well as the exact position on the earth's surface at which it can be seen, but a description of the method would take us far outside the scope of this book.

The Chaldean Saros, is a cyclical period consisting of 6585 days (i.e., 18 years and 11 or 10 days—according as there are 4 or 5 leap years in that period), during which eclipses repeat themselves in exactly the same order as in the preceding period of 6585 days. The Chaldeans discovered that the number 6585 contains almost an exact number of synodic periods of the nodes ($\frac{6585}{346{\cdot}62}$ = 28·998, which is practically 29), as well as an integral number of synodic months ($\frac{6585}{29{\cdot}53}$ = 222·994, which is practically) 223). Therefore after every cycle of 6585 days, both the moon

[1] Yesod Olam, iii, 17.

and its nodes return to exactly the same position in relation to the sun, and the eclipses, therefore, recur. As the Babylonians kept careful records of all their eclipses (Ptolemy possessed a Babylonian record of eclipses going back to 747 B.C.), they were able to use this period—called the *Saros* (which means repetition) to predict the recurrence of eclipses.

The Hebrew Prophets were also apparently able to foretell the occurrences of eclipses, and in view of the ignorance and credulity of the masses, threatened them with coming disaster —such as darkening of the sun, moon, or stars—as a punishment for their sins. Thus Isaiah (xiii., 10, 11) says: " For the stars of heaven and the constellations thereof shall not give their light; the sun shall be darkened in his going forth and the moon shall not cause her light to shine. And I will punish the world for their evil "

כי כוכבי השמים וכסיליהם לא יהלו אורם, חשך השמש בצאתו וירח לא יגיה אורו ופקדתי על תבל רעה. Compare also Amos viii., 9.

The Saros method does not, however, enable one to calculate the character and details of an eclipse—or the place on the earth's surface where such an eclipse will be visible; these can be computed by strictly accurate mathematical methods, which are discussed at length in books on Spherical Astronomy, but which we have sufficiently indicated in the foregoing pages.

Eclipses and Jewish Chronology. It is, as we have just said, possible to forecast by calculation the date, the exact time and duration of solar and lunar eclipses and the places on the earth's surface where they would occur. It is equally possible to calculate backwards for any number of centuries and in this way establish with certainty the date of some event which is known to have occurred at or about the time of an eclipse of recorded character. In Jewish Chronology eclipses do not play the same important role as they do in the Chronology of other ancient peoples, because throughout the ancient Jewish literature there is no description given of any indisputable eclipses. There are, however, a considerable number of passages in the Bible which obviously allude to these phenomena, and some of these have been thought to refer to eclipses that occurred about the time of the

events related in those passages. We shall take the following examples: -

(1) **The "darkness" at the time God made his Covenant with Abraham** (Gen. xv.) ברית בין הבתרים :

" And when the sun was going down, a deep sleep fell upon Abram, and lo, an horror of great darkness והנה אימה חשכה גדולה fell upon him " (Gen. xv., 12).

Calculation shows that on the 8th October, 1764 B.C. there was an almost total eclipse of the sun ($\frac{23}{24}$ of the sun's disc having been obscured) in Canaan, about 5 hours before sunset. There were other eclipses in that land in the same century, but none of these fulfilled the conditions given in the Bible. Thus, the " darkness " having occurred " when the sun was going down ", the time of the day must have been in the *afternoon.* Therefore all eclipses which may have occurred *before noon* must be excluded. Secondly, according to Mahler[1], all eclipses occurring too late in the afternoon must be excluded because—he points out—Abraham had time between the " darkness " and sunset to kill and cut up into pieces a calf, a goat, a ram and two birds (Ibid. 9 and 10). Lastly, the " darkness " must have been sufficiently intense for stars to have been visible (Ibid. 5), and therefore all eclipses which did not obscure a large enough portion of the sun's disc must also be excluded.

It seems to the author that Mahler's arguments, though highly ingenious, are somewhat forced. A careful reading of Gen.xv. shows that when Abraham was told to look towards heaven and count the stars, the " darkness " in question had not yet set in, so that this incident must have occurred on some previous evening. The same applies to the cutting up of the animals.

(2) **The Plague of darkness.** (Exod. x., 22-23).

Here again Mahler ascribes the " darkness " to the natural phenomenon of a solar eclipse which occurred in Egypt on Thursday, 13th March, 1335 B.C., *thus fixing the exodus, which occurred 14 days later, at 27 March, 1335 B.C.*

Mahler argues that the Biblical description of the darkness makes it clear that if it was due to an eclipse, the latter must have

[1] Handb. d. Jüd. Chronol. Leipzig 1916 pp. 230 et Seq.

been a total one, and as the exodus occurred on Thursday, 15th Nisan[1] (ניסן שבו יצאו ישראל ממצרים... בחמשה עשר יצאו... ואותו היום חמישי בשבת היה) and the plague of darkness took place on the 1st Nisan, that day must also have been a Thursday. Hence, if there was any eclipse in the 13th or 14th centuries B.C. which was total, and fell on a Thursday corresponding to the 1st Nisan, the year of that eclipse would fix the year and day of the month of the exodus. There was only one eclipse which fulfilled these conditions, and that was the one occurring on Thursday, 13th March, 1335 B.C. Hence the exodus took place on 27th March, 1335 B.C.

Mahler overcomes certain difficulties as follows:-

(a) According to the Biblical version the "darkness" lasted 3 days, which would therefore appear to militate against its having been a solar eclipse, which is, of course, of short duration. But the difficulty vanishes if we accept the suggestion that the punctuation of verses 22 and 23 as given in the current editions of the Bible is erroneous. The verses as printed read:-

> verse 22. ...there was a thick darkness in all the land of Egypt for 3 days.
>
> verse 23. They saw not one another, neither rose any from his place for 3 days.

Mahler's suggestion is that the last 3 words of verse 22 really belong to verse 23 as follows:-

> verse 22. ...there was a thick darkness in all the land of Egypt.
>
> verse 23. For 3 days they saw not one another, neither rose any from his place for 3 days.

The eclipse, therefore, really lasted only a short while, but it had such a moral effect upon the Egyptians that for 3 days they were afraid to move.

(b) The other difficulty is: If the darkness was attributable to a non-miraculous agency, how can one account for the statement in verse 23 that "all the children of Israel had light in their dwellings"? But as Mahler points out, the eclipse was total only in the lower Nile Delta, but in Goshen—where the Jews lived

[1] Sabbath 87b.

—the darkness was not complete, since in that place $\frac{1}{12}$ of the sun's disc was uneclipsed.

(3) **The date of Amos** nearly coincides with a total eclipse which occurred in Jerusalem shortly after noon on February 9th, 784 B.C., and hence one may say that his prophecy (Amos viii., 9) that God " will cause the sun to go down at noon, and will darken the earth in the clear day " (והבאתי השמש בצהרים והחשכתי לארץ ביום אור) refers to that eclipse.

The Talmud contains nothing of scientific value regarding the nature and prediction of eclipses. It attributes them to the following causes: " Failure to deliver a funeral oration for a departed Chief Judge; rape; unnatural sexual offences; simultaneous murdering of two brothers; libel;(כותבי פלסתר);perjury, etc." [1]

Rashi comments that he cannot explain the relationship between cause and effect in these cases. Israeli[2], however, suggests that these causes are merely cryptic ways of expressing the fact that eclipses can only occur during *conjunction* or *opposition*, and when the moon is at the *node*. Thus, the death of a chief judge refers to a new moon or conjunction, since the moon is sometimes spoken of as a representative of the King (the sun) and during a conjunction the moon is said to die. Failure to declare public mourning is meant to indicate that the moon by failing during a conjunction to get out of the sun's way (the node), thus causing a solar eclipse, has behaved like a King's representative rebelling against the King, thus forfeiting the right to have respect paid to him after his death. An ingenious but utterly unwarranted interpretation!

In common with others, the Rabbis regarded eclipses as bad omens. Thus, notwithstanding the teaching of Jeremiah (x., 2 and 5) not to be " dismayed at the signs of the heavens.........for they cannot do evil, neither is it in them to do good ", (מאותות השמים אל תחתו, כי יחתו הגוים מהמה... כי לא ירעו וגם היטיב אין אותם), some of the Rabbis stated that an eclipse of the sun is bad for the heathens and an eclipse of the moon is bad for the Jews, because the heathens count time by the sun and the Jews do so by the moon.[3]

Use of Eclipses for the Determination of the Average length of a Synodic Month. As a synodic month is the interval between 2 successive conjunctions, or—which is the same thing

[1] Succah 29a. [2] Yesod Olam, iii, 17. [3] Succah 29a.

—between 2 successive oppositions, and as a lunar eclipse can only occur when the moon is in opposition, it is obvious that measurement of the exact intervals between the middle of two lunar eclipses, sufficiently long removed from each other, divided by the number of lunations—as measured by the number of full moons—between them, gives the average length of a single lunation or synodic month. Hipparchus, by measuring the exact interval between a lunar eclipse recorded by the Babylonians and one observed by himself, found that 4267 full moons (and therefore lunations) occurred in the interval of 12607 days 1 hour, i.e., in 3,024,169 hours. Therefore:

$$\textit{Mean duration of Synodic Month} = \frac{3{,}024{,}169 \text{ hours}}{4267}$$

$$= 29d.\ 12h.\ 44m.\ 3{\cdot}3s$$
$$= 29{\cdot}53 \text{ days.}$$

The Theory of the Moon. The ancient astronomers recognised the much greater complexity of the moon's motion as compared with that of the sun. Thus they found from observation of eclipses that the length of a synodic month is not constant, but varies within a few hours on either side of a mean of 29d. 12h. 44m. 3⅓s. In addition, there are very many irregularities in the moon's motion in various parts of her course, irregularities which we now know as *Variation*, *Evection*, *Annual Equation*, *Secular Acceleration* and numerous others. With the nature of these irregularities or *perturbations* we need not concern ourselves at the moment; they are all satisfactorily explained by Newton's Law of Gravitation, and by the fact that the lunar orbit, is, in accordance with Kepler's laws, an ellipse of eccentricity varying between 0·044 and 0·066 in different parts of its course—with a mean value of 0·055 (see p. 154). The ancient astronomers, however, having failed to explain these various irregularities by means of an eccentric circle alone—as they did in the case of the sun—and being obsessed with the idea that heavenly bodies whose courses were "ordained by God" could have no other than *perfect* movements, which to their minds was synonymous with *Circular* movements, represented the moon's motion by an ingenious device of a *Combination of Circular Motions*. This theory, which prevailed up to the time of Kepler (16th century), assumed the centre of the moon to be moving uniformly round the circumference of a small circle called the *epicycle* (גלגל קטן שאינו מקיף את העולם[1] or גלגל הקפה[2]), whose centre is simultaneously moving round the earth along the circumference of another, large, circle called the *deferent* גלגל גדול[1])

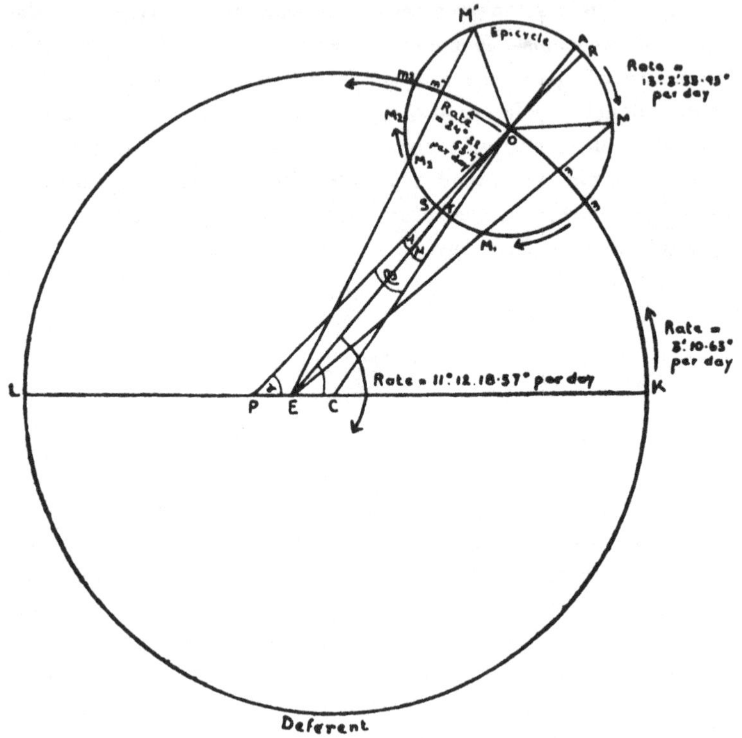

Fig. 45

(גלגל הנושא[2] or המקיף את העולם), in which the earth is eccentrically situated at E (Fig. 45)

Ptolemy's Epicycle Theory. In Fig.45, C represents the centre of the deferent (as well as of the Celestial Sphere); E represents the centre of the earth, situated at a fixed distance east of C; O is the centre of the epicycle and M′, M″, M_1, M_2 are 4 positions of the centre of the moon on the circumference of the epicycle. The distance EC is such that the ratio $\frac{EC}{CO}$ or $\frac{EC}{CK} = \frac{619}{2981} =$ say $\frac{1}{5}$.

The line ETOA, joining the centre of the earth with the centre of the epicycle and cutting the epicycle in the diameter TA,

[1] Kidd. Hach. xiv. 1. [2] Yesod Olam, iii. 8.

is called the *Apse line* (קו המישור), whilst the line EM′ or EM″ etc., joining the centre of the earth with the centre of the moon at any of its positions on the circumference of the epicycle is called the *indicator* (קו המורה).

Prosneusis. To complete his theory of lunar motion, Ptolemy imagined the existence of another point, P, on the continuation of the line CE, such that PE = EC, and by joining this point P with the centre of the epicycle, O, so as to cut the epicycle in the diameter SR, he postulated an oscillation of the epicycle, to and fro, between the points A and R, i.e., to an extent varying with the size of the angle AOR or POE. This to-and-fro oscillation is called *Prosneusis* (גליוה) and the point P is called the *Prosneusis point* (נקדת הגליוה)[1] The size of the prosneusis angle POE depends upon that of the angle OEK (see p. 144).

The Mean and True Position of the Moon. The position O of the centre of the epicycle on the circumference of the deferent at any moment is called the moon's *mean position* (מקום אמצע הירח)[2] and the moment at which O has the same longitude as the sun's mean longitude is called the *mean conjunction* or *Molad* (מולד). The point m at which the indicator EM cuts the circumference of the deferent is called the moon's *true position* (מקום הירח האמתי)[3] and the moment at which m has the same longitude as the sun's true longitude is called the *true conjunction*, or true new moon. The angle AOM, measured by the arc AM of the epicycle, representing the distance between the position of Apogee (A) and that of the Moon (M) on the epicycle, is called the moon's *Anomaly* (מסלול)[4] whilst the arc Om, which is the difference between the mean and true position of the moon, is called the *Equation of the Centre* (מנת המסלול).[5]

As the centre O of the epicycle moves uniformly round the deferent, the mean conjunctions follow one another uniformly, and therefore the mean synodic month is of constant length (viz., 29d. 12h. 44m. $3\frac{1}{3}$s.). But the duration of a true synodic month is variable, because it depends upon the position of M in relation to the sun and O. If, when O has the same longitude as the sun, M also happens to have the same longitude (e.g., when M is at

[1] Yesod Olam, iii. 15. [2] Kidd. Hach. xv. 1. [3] Ibid.
[4] Ibid xiv. 3. [5] Ibid xv. 3.

A or T, which is the case when the moon is at apogee or perigee), then the mean and true conjunctions coincide. If, however, the moon M happens to be at any other point, then :-

(a) If it is either at M′ or M_1, i.e., on the west side of the apse line, then the point O has already anticipated in longitude the point M′ or M_1, and therefore the mean conjunction has taken place before the true conjunction—i.e., the mean synodic month is shorter than the true synodic month.

(b) If the moon happens to be at M″ or M_2, then obviously the mean synodic month is longer than the true synodic month.

(c) The greatest difference in time between the true and mean synodic months is when the moon, at a true conjunction, happens to be at m_1 or m_2, i.e., the points of intersection of the epicycle with the deferent. When conjunction occurs at m_1 the true synodic month has the longest duration; and when it occurs at m_2 the true synodic month is shortest.

By knowing a few ratios as well as a few rates of movement, the position of the moon as well as its visibility at any moment can be calculated.

These are given in the following table :-

Numerical Data regarding the Motion of the Moon.

(a) *Ratios* (given by Ptolemy, and also by Israeli, but not by Maimonides)

(i) Eccentricity of Deferent, i.e., $\frac{EC}{CK} = \frac{619}{2981} = 0{\cdot}208$ = say $\frac{1}{5}$.

(ii) $\frac{\text{Distance of Perigee,}}{\text{Distance of Apogee}}$ i.e. $\frac{LE}{EK} = \frac{2981-619}{2981+619} = \frac{2362}{3600}$ $= 0{\cdot}6561 =$ say $\frac{2}{3}$.

(iii) $\frac{\text{Radius of Epicycle,}}{\text{Radius of Deferent}}$ i.e. $\frac{OM}{CO} = \frac{314}{2981} = 0{\cdot}1053 =$ say $\frac{1}{10}$.

(iv) $\frac{\text{Radius of Epicycle}}{\text{Mean distance between centre of Earth and that of Epicycle,}}$

i.e. $\frac{OM}{\text{mean } EO} = \frac{314}{3564} = 0{\cdot}088 =$ say $\frac{1}{11}$.

(b) *Rates of Motion*, (The direct motions are positive and indicated by + ; the retrograde motions are negative and indicated by –.).

(i) *Moon round Epicycle*: Retrograde, or East to West motion in the upper half of the circle and from West to East in the lower half: – 13°3′53·93″ per day[1] (=– 32′40″ per minute or about 33″ per second).

(ii) *Deferent.*

(α) Centre of Deferent (C) round centre of Ecliptic: retrograde motion, –11°12′18·37″ per day

(β) Centre of Epicycle (O) round circumference of Deferent: Direct West to East motion, 24°22′53·40″ per day.

(γ) Therefore, net movement of centre of Epicycle round Deferent: Direct, West to East motion (β – α)=13°10′35·03″[2] (=33′ per minute or 33″ per second).

(iii) *Nodes* round Ecliptic: Retrograde East to West motion, – 3′10·63″ per day. Maimonides[3] gives it as 3′11″ per day, but more accurately as 5°17′43″ in 100 days, which means a rate of 3′10·63″ per day.

(iv) *Sun*: Direct West to East motion.

(α) Tropical (i.e., with reference to Equinox) 0°59′8·33″ per day[4]

(β) Sidereal (i.e., with reference to fixed star) 0°59′8·19″ per day.

The Talmud nowhere refers to the Epicycle and Deferent or to their rates of movement.

Different kinds of Month. A month is the period during which the moon returns to a given starting point in the heavens. Hence it has different lengths according as the starting point is

[1] Kidd. Hach. xiv. 3. [2] Ibid, xiv, 2. [3] Ibid, xvi, 3. [4] Ibid, xii. 1.

stationary (like a fixed star) or moving; and if the latter, according to its rate and direction of motion. We thus get the following table:-

Name of Month	Starting Point		Duration of Month
	Name	Rate & Directⁿ of Motion	
Sidereal	Fixed Star	Nil	27d. 7h. 43m. 11s. (Making the mean daily rate 13°10′34·89″.)
Synodic	Centre of Sun	59′8″·33″ per day (same)	$\frac{360^\circ}{\underbrace{12^\circ 11' 26{\cdot}7''}_{\text{Daily Elongation}}}$ = 29d.12h.44m.2·63s.
Tropical[1]	Equinox	4″ per month (opposite)	$\frac{359^\circ 59' 56''}{360^\circ}$ × 27d. 7h. 43m. 11s. = 27d 7h. 43m. 4s. (Making the mean daily rate 13°10′35·03″)
Draconic or Nodal	Node	3′10·63″ per day (opposite)	$\frac{360^\circ}{13^\circ 13' 4542''}$ = 27d. 5h. 5m. 40s.
Anomalistic[2]	Lunar Apogee	6′41·10″ per day (same)	$\frac{360^\circ}{13^\circ 3' 53{\cdot}9\,3''}$ = 27d. 13h. 18m. 35s.

The denominators in the last column in the case of the synodic, draconic, and anomalistic months, is the algebraic sum of the daily rates of motion of the moon (in relation to the equinox) and of the starting point in question. Thus $12^\circ 11' 26{\cdot}7'' = 13^\circ 10' 35{\cdot}03'' - 0^\circ 59' 8{\cdot}33''$; $13^\circ 13' 45{\cdot}52'' = 13^\circ 10' 35{\cdot}03'' + 3' 10{\cdot}63''$; $13^\circ 3' 53{\cdot}93'' = 13^\circ 10' 35{\cdot}03'' - 6' 41{\cdot}10''$.

NOTE: We know that, for gravitational reasons, the synodic month is gradually diminishing at the rate of about 0·2 sec. per 1000 years. As its present value is 29d. 12h. 44m. 2·8s., its value at the time of Maimonides must have been about 29d. 12h. 44m. 3s. Hence the value given by Maimonides for the moon's daily rate of motion viz. 13°10′35·03″ is too high. This is all the more remarkable in view of the fact that his value of 59′8·33″ for the mean daily motion of the sun is so amazingly exact that it makes the tropical year only 0·2 sec. higher than its true value at that time, although the value given by Ptolemy is somewhat higher, and that given by Al Battani somewhat lower, than that given by himself.

Moreover, he himself adopts as the basis of the fixed Jewish Calendar a mean synodic month of 29d. 12h. 44m. 3⅓s.[3], which is also the value given by Ptolemy. Such a value for the synodic month makes the mean daily elongation 12°11′26·69″ and the mean daily rate of lunar motion 13°10′35·02″.

1 In the Ptolemaic system, a tropical month is the time of a complete revolution of the centre of the epicycle round the Deferent.

In the Ptolemaic system, an Anomalistic month is the period of revolution of the moon round the epicycle. 3Kidd. Hach. vi, 3.

The Double Elongation (מרחק כפול)[1] Let the moon's mean position (i.e., point O in Fig. 45), its apogee (point A) and the sun, have the same longitude (corresponding say to the point K). After 24 hours, O will have moved from *west to east* over a net arc of 13°10′35·03″ (b. ii. γ), and the sun will have moved in the same direction over an arc equal to 0°59′8·33″.

Therefore: *Mean elongation of moon after 24 hours* = 13°10′35·03″ − 0°59′8·33″ = 12°11′26·7″.

But during the same interval, the movement of the centre C of the deferent round the centre E of the earth is in an opposite direction, from *east to west* to an extent of 11°12′18·37″ (see b. ii.α) carrying apogee with it, so that:-

Distance between moon and apogee after 24 hours = 13°10′35·03″ + 11°12′18·37″ = 24°22′53·40″.

Hence we have the *result which is universally true that:- Distance of Mean Position of Moon from Apogee at any time* (angle OEK) *is equal to twice the elongation at that moment.*

Maimonides calls the elongation מרחק and therefore he calls the angle OEK מרחק כפול " double elongation ", which is a most important quantity for calculating the position of the moon (see Chapter XIII).

Ptolemy's theory of the moon on which Maimonides's *Hilchoth Kiddush Hachodesh* is of course based, is sufficiently accurate to

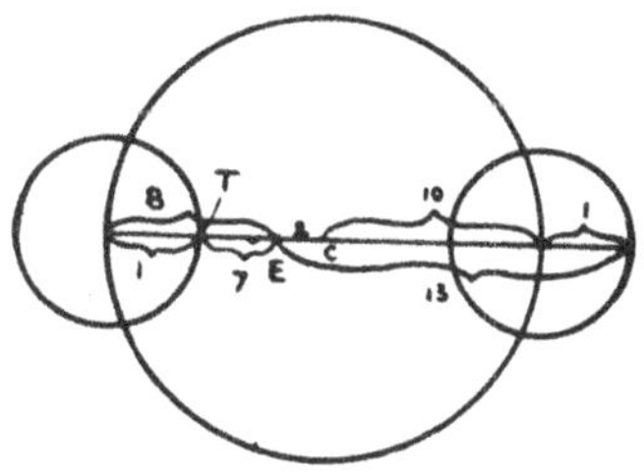

Fig. 46

calculate the moon's true position at any moment (see next chapter). The great flaw in it, however, lies in the fact that when the moon is in the position A (Fig. 46), its distance from the earth, E, is 13,

[1] Kidd. Hach. xv. 1.

whilst when it is in position T, its distance from the earth is no more than 7, and therefore, on this theory, the maximum and minimum apparent diameters of the moon should be in the proportion of 13 : 7, i.e., nearly 2 : 1, when, in point of fact, the ratio is no more than about 67 : 59 or 1·136 : 1. We know now that the moon's orbit is very nearly an ellipse whose eccentricity is about $\frac{1}{18}$ (see p. 154).

CHAPTER XIII.

The Position of the Moon in the Heavens.

כלם נכחים למבין וישרים למוצאי דעת (משלי ח' ט'.)

"They are all plain to him that understandeth, and right to them that find knowledge" (Proverbs, viii, 9)

To determine whether the moon is visible at any moment, and to ascertain the exact moment of a true conjunction, it is necessary to know the moon's true position (i.e., its true latitude and longitude) at that moment.

(A) Longitude.

(a) *First find the moon's mean longitude* אמצע הירח[1] (i.e., the longitude of the point O in Fig. 45). As the mean increase in lunar longitude is at the rate of 13°10′35·03″ a day[1], or at the rate of about 32′56·3″ an hour or 32·93″ a minute or 0·55″ a second, and its mean longitude at the epoch (sunset of 2nd Nisan, 4938 W.E.) was 31°14′43″ (XV., 4), therefore the mean longitude of the moon at the moment in question is 31°14′43″ ± n × 13°10′35·03″—where *n* is the interval in days and fractions of a day after, or before, the epoch. Thus if n = 74d. 3h. 25m. 7s. we have:-

$$\textit{Required mean longitude} = 31°14'43'' + \underbrace{74 \times 13°10'35{\cdot}03''}_{255°3'12''}$$

$$+ \underbrace{3 \times 32'56{\cdot}3''}_{1°38'49''} + \underbrace{25 \times 32{\cdot}93''}_{13'43''} + \underbrace{7 \times 0{\cdot}55''}_{4''} = 288°10'31''.$$

The problem with which the Talmudic Rabbis were confronted, however, was to ascertain the moon's mean longitude at *sunset* or at a given interval after sunset on a given day. This, therefore, is the way that Maimonides deals with the subject (XV. 5, 6). It is obvious that we must first of all compute (in the manner described in Chapter IX.) the exact moment of sunset on the given day. If that is later than the moment of sunset at the epoch, we must add that difference (multiplied by 32′56″ per

[1] Kidd. Hach. xiv. 1.

hour, etc.) to $n \times 13°10'35{\cdot}03''$, where n is the number of complete days elapsed since the epoch; if it is earlier, the difference must be subtracted, and if it is the same, there is nothing to be added or subtracted.

The time of sunset is calculated by first finding the sun's longitude on the day in question (which is equal to the longitude at the epoch $\pm\ n \times 59'8{\cdot}33''$), and using the longitude λ thus found to compute the sun's declination on the same day by means of the equation $\text{Sin}\ \delta = 0{\cdot}3987\ \text{Sin}\ \lambda$ (p. 84). The time of sunset $\frac{SD}{15}$ (Fig. 36), is then found from the equation $\text{Cos SD} = -\tan\phi\tan\delta$ (p. 90).

Example. To find the time of sunset in Jerusalem 29 days after the Epoch ($\phi = 32°$).

We have, $\lambda = 7°3'32'' + 29 \times 59'8{\cdot}33'' = 35°38'33''$[1]

$$\therefore\quad \text{Sin}\ \delta = 0{\cdot}3987\ \text{Sin}\ 35°38'33'' = 0{\cdot}3987 \times {\cdot}5827 = {\cdot}2323$$

whence $\delta = 13°26'$

$$\therefore\quad \text{Cos SD} = -\tan 32° \tan 13°26' = -{\cdot}6249 \times {\cdot}2388 = -{\cdot}1492$$

$$\therefore\quad SD = 98°35', \text{ and } \frac{SD}{15} = 6\text{h. } 34\text{m. } 20\text{s. p.m.},$$

i.e., $6.34\frac{1}{3}$ p.m.

Now, at epoch, $\text{Sin}\ \delta = 0{\cdot}3987\ \text{Sin}\ 7°3'32'' = 0{\cdot}3987 \times 0{\cdot}1230$
$= 0{\cdot}0490$, whence $\delta = 2°48'$.

$$\therefore\quad \text{Cos SD} = -{\cdot}6249 \tan 2°48' = -{\cdot}6249 \times 0{\cdot}0489 = -\ 0{\cdot}0306$$

$$\therefore\quad SD = \text{about } 91°45', \text{ and } \frac{SD}{15} = 6\text{h. } 7\text{m., i.e., } 6.7 \text{ p.m.}$$

$\therefore$ Difference in times of sunsets on the two dates $= 27\frac{1}{3}$ mins., during which the increase in moon's longitude is about 15′.

Hence we get: *Mean longitude of moon at sunset of the day in question* $= 31°14'43'' + 29 \times 13°10'35{\cdot}03'' + 15' = 53°36'39''$, which is the figure given by Maimonides.[2]

[1] Kidd. Hach. xv. 8. [2] Ibid.

Working in this way we get the following Table:-[1]

λ	δ	Sunset	Additional Arc	
			Calculation	Maimonides
7°3′32″ (Epoch)	2°48′	6h. 7m.	— —	—
15°	5°56′	6h. 15m.	8 × 32·93″ = 4′24″	0
60°	20°12′	6h. 53m.	46 × 32·93″ = 25′15″	15′
120°	20°12′	6h. 53m.	46 × 32·93″ = 25′15″	15′ (30′)
165°	5°56′	6h. 15m.	8 × 32·93″ = 4′24″	15′
195°	− 5°56′	5h. 45m.	− 22 × 32·93″ = − 12′ 4″	0
240°	− 20°12′	5h. 7m.	− 1 × 32′56″ = − 32′56″	− 15′
300°	− 20°12′	5h. 7m.	− 1 × 32′56″ = − 32′56″	− 30′
345°	− 5°56′	5h. 45m.	− 22 × 32·93″ = − 12′ 4″	− 15′

It will be seen, therefore, that Maimonides who aims at giving values of the additional arc correct to the nearest 15′, gives correct values for dates when the solar longitude is 15°, 300°, or 345°, but his values are not in agreement with calculation for the other solar longitudes. In the case of λ = 120° the Frankfurt edition of *Hilchoth Kiddush Hachodesh* (1720), edited by R. Jonathan ben Joseph of Rosnoy, gives the value of the additional arc as 30′ instead of 15′, which is in agreement with calculation. The same value is given in 14th and 15th century manuscripts. (British Museum and Mr. D. S. Sassoon's Collections.) The Vatican Manuscript gives the value of the additional arc for λ = 120°, as 30′, but gives no value for λ = 165°.

(b) *Having found the moon's mean longitude* (i.e., the position of the centre O of the epicycle) we must *convert it into true longitude*[2] מקום הירח האמתי (i.e., the longitude of the point m in which the line joining the centre of the earth E with that of the moon M on the epicycle cuts the deferent). This is done by finding the value of the arc Om, and either subtracting it from, or adding it to, the mean longitude of O—according as M is at M′ or M″ (i.e., according as the anomaly is less or greater than 180°). *Hence our problem is to find the size of the Arc Om* or, which is practically the same thing, the *angle OEM*. This is done as follows:-

(a) First find the size of the arc AM of the epicycle, consisting of:

(1) Arc AR—the prosneusis arc—(which = angle POE, the prosneusis angle).

[1] Kidd. Hach. xiv. 5. [2] Kidd. Hach. xv. 1.

and (2) Arc RM—the moon's mean anomaly, i.e., its mean distance from its own apogee.

(1) *Prosneusis Arc AR, or angle POE.* This is found as follows:

(i) First solve triangle ECO, of which EC and CO are known (= ·208 and 1 respectively) and angle OEC = twice the elongation at the moment, which is also known, if elongation at epoch is known—since elongation increases at the rate of 12°11′26·70″ per day, or 30′28·6″ per hour, or 30·5″ per second. Thus from the well-known trigonometrical relation $\frac{\text{Sin A}}{\text{a}} = \frac{\text{Sin B}}{\text{b}}$ which holds good for all triangles we have $\frac{\text{Sin EOC}}{\cdot 208} = \frac{\text{Sin OEC}}{1}$, whence Sin EOC = ·208 Sin OEC. Thus, to pursue Maimonides' problem (XV. 8):

Mean lunar longitude at sunset, 29 days after epoch	= 53°36′39″
Mean solar longitude at same time	= 35°38′33″
∴ Mean Elongation at that moment	= 17°58′ 6″

∴ OEC = 2 × 17°58′6″ = 35°56′12″

∴ Sin EOC = ·208 Sin 35°56′12″ = ·208 × ·5869 = ·1221

∴ EOC = 7°0′43″

Also angle ECO = 180° − (35°56′12″ + 7°0′43″) = 137°3′5″

[Further, the side EO can also be computed (see p. 145).]

(ii) Now solve triangle PCO, of which PC and CO are known (= 1238 and 2981 respectively), and contained angle PCO (or ECO) has just been found. Thus by the well-known relation $\frac{a-b}{a+b} = \frac{\tan\frac{1}{2}(A-B)}{\tan\frac{1}{2}(A+B)}$ which is universally true, we have :-

$$\frac{CO - PC}{CO + PC} = \frac{\tan\frac{1}{2}(CPO - COP)}{\tan\frac{1}{2}(CPO + COP)}$$

$$\text{i.e., } \frac{2981 - 1238}{2981 + 1238} \text{ or } 0{\cdot}4131 = \frac{\tan\frac{1}{2}(CPO - COP)}{\tan\frac{1}{2}(CPO + COP)}$$

But CPO + COP = 180° − PCO, is known.

∴ CPO − COP is ascertained.

∴ By subtraction COP is evaluated.

But COP = COE + EOP, and COE, or EOC, is known.

$\therefore$ EOP, or POE, the prosneusis angle, is ascertained.

Thus, in the case of our example from Maimonides (xv. 8)

$$\text{CPO} + \text{COP} = 180° - \text{PCO} = 180° - \text{ECO} = 180° - 137°3'5''$$
$$= 42°56'55''$$

$$\therefore \quad \tan \tfrac{1}{2}(\text{CPO} - \text{COP}) = \cdot 4131 \tan 21°28'28''$$
$$= \cdot 4131 \times \cdot 3934 = \cdot 1625$$

$$\therefore \quad \frac{\text{CPO} - \text{COP}}{2} = 9°13'46'', \text{ whence CPO} - \text{COP} = 18°27'36''$$

$$\text{But} \qquad \text{CPO} + \text{COP} = 42°56'55''$$

$$\therefore \quad \text{by subtraction COP} = \frac{24°29'19''}{2} = 12°14'40''$$

$$\therefore \quad \text{POE} = \text{COP} - \text{EOC} = 12°14'40'' - 7°0'43'' = 5°13'57''$$

$= 5°$, to the nearest degree (אין מקפידין על החלקים במסלול).

We thus see that the size of AR, or POE, depends entirely upon the angle EOC (מרחק כפול) from which it can be calculated at any moment.

(2) *Mean anomaly, or Arc RM.* The value of this arc at any moment is known from its value at the epoch, since mean anomaly increases at the rate of 13°3′53·93″ a day, or 32′40″ an hour. Here, it is not necessary to allow for the difference in the times of sunset on the two dates, because the longest or shortest day in Jerusalem differs from an equinoctial day by about 1 hour only (see p. 91), during which time mean anomaly increases by no more than 32′40″—which when converted into arc of longitude, will not affect the arc Om by more than about 3′ (since $\frac{\text{OM}}{\text{EO}} = \frac{1}{11}$). On all other days, the difference in Om corresponding to the difference in the times of sunset will amount to less than 3′, and may, therefore, be ignored.

Hence, in the case considered by Maimonides, we have:

Mean Anomaly, 29 days after the epoch = 84°28′42″ [1] + 29 × 13°3′53·93″
= 103°21′46″.

The arc AR must always be added to RM. For, although by the time the centre of the epicycle has travelled from K to O, the moon has travelled from R to M, the epicycle itself has in the meantime swung from A to R, and, therefore, the total arc covered by the moon is AM.

[1] i.e. Mean Annomaly at Epoch (Kidd. Hach. xiv. 4).

We thus have: *AM* (*i.e. true Anomaly*, מסלול הנכון)
= *RM* (*i.e. mean Anomaly*, אמצע המסלול)
+ *AR* (*i.e. Prosneusis Arc*, תוספת המסלול)

Therefore, in our case, the *True Anomaly* = 103°21′46″ + 5° = 108°22′ to the nearest minute, which Maimonides, probably through a slip of the pen, approximates to 108°21′.

Maimonides, who calls the prosneusis arc AR תוספת המסלול, (i.e., anomaly excess),[1] gives its value to the nearest 1° (see Table) for different values of the double elongation OEC, from 5° to 63°.

Double Elongation (angle OEC) (מרחק כפול)	Size of Arc AR (Prosneusis) (תוספת המסלול)		
	Calculation		Maimonides
	Range	Mean	
0°– 5°	0°– 0°44′	0°22′	0°
6°–11°	0°52′–1°37′	1°11′	1°
12°–18°	1°45′–2°38′	2°11′	2°
19°–24°	2°40′–3°33′	3° 7′	3°
25°–31°	3°35′–4°32′	4° 4′	4°
32°–38°	4°33′–5°32′	5° 3′	5°
39°–45°	5°33′–6°31′	6° 2′	6°
46°–51°	6°32′–7°22′	6°57′	7°
52°–59°	7°23′–8°27′	7°55′	8°
60°–63° [2]	8°28′–8°58′	8°43′	9°

He does not concern himself with values outside these limits because, as we shall see later, when the elongation is less than $2\frac{1}{2}$ (i.e., $\frac{5}{2}$), the moon is too near the sun to be visible, and when the elongation exceeds 31° (i.e., about $\frac{63}{2}$), it is so far from the sun that it cannot fail to be visible everywhere. It is only when the elongation is between these limits that the Calendar Council had to calculate the exact position of the moon in the heavens in order to ascertain whether, at the moment that the moon is alleged to have been seen by the witnesses, the difference in altitude between itself and the sun was sufficiently great to render the moon visible (see Chapter XV).

(*β*) Having found AM, which is the same as the angle AOM, which = OME + OEM, and knowing the values of OM and EO (since EO has been found under *a* i.), we can now solve the triangle EOM (again by using the relationship $\frac{a-b}{a+b} = \frac{\tan \frac{1}{2}(A-B)}{\tan \frac{1}{2}(A+B)}$

[1] K'dd Hach., xv. 3. [2] The Vatican MS. gives משׁשים מעלות עד שלש מאות וששים מעלות, i.e. 60°–360°, which is obviously a lapsus calami; it should have been עד שלש וששים.

and get the value of the angle OEM—which is practically the same as the arc Om.

$$\text{Thus, } \frac{\text{Sin ECO}}{\text{EO}} = \frac{\text{Sin EOC}}{\text{EC}}$$

$$\therefore \quad \text{EO} = \frac{\text{EC Sin ECO}}{\text{Sin EOC}} = \frac{619 \text{ Sin } 137^\circ 4' 48''}{\text{Sin } 7^\circ 0' 43''} = \frac{619 \times \cdot 6811}{\cdot 1221} = 3453$$

It is obvious that EO varies with the size of the angle OEC ("double elongation") from 3600, i.e., 2981 + 619 (when OEC = 0°) to 2362, i.e., 2981 − 619 (when OEC = 180°). If, however, we take the mean length of EO between OEC = 5°, and OEC = 63° (the practicable limits), say when OEC = 34°, it will be found to be about 3470.

Therefore, in triangle EOM, we have EO = 3470 and OM = 314.

$$\therefore \quad \frac{3470 - 314}{3470 + 314} = \frac{\tan \frac{1}{2}(\text{OME} - \text{OEM})}{\tan \frac{1}{2}(\text{OME} + \text{OEM})} \text{ i.e., } \tan \tfrac{1}{2}(\text{OME} - \text{OEM})$$

$$= \cdot 8340 \tan \tfrac{1}{2}(\text{OME} + \text{OEM})$$

Thus, in the case of our problem from Maimonides (xv. 8)

$$\text{OME} + \text{OEM} = \text{AM} = \text{AR} + \text{RM} = 108^\circ 21' \text{ (p. 144)}$$

$$\therefore \quad \tan \tfrac{1}{2}(\text{OME} - \text{OEM}) = \cdot 8040 \tan 54^\circ 10' 30'' = \cdot 8040 \times 1 \cdot 3855$$

$$= 1 \cdot 1553$$

$$\therefore \quad \frac{\text{OME} - \text{OEM}}{2} = 49^\circ 7' 17''$$

$$\text{But } \frac{\text{OME} + \text{OEM}}{2} = 54^\circ 10' 30''$$

$$\therefore \text{OEM} = 5^\circ 3' 13'', \text{ say } 5^\circ 3'.$$

As the true longitude of the moon is less than 180°, therefore Om has, as in the case of the sun (p. 115), to be subtracted from it.

Therefore **True Longitude** = 53°36′39″ − 5°3′

= 48°34′ to the nearest minute.

Maimonides apparently calculating for a mean OEC of 30° (making EO = 3495), gets OEM = 5°1′, and therefore obtains the value of the true longitude as 48°36′.

The arc Om (or angle OEM) is called the *Equation of the Centre*. Maimonides[1] calls it the *Quota of the Anomaly* (מנת המסלול), because its value depends upon the size of the true anomoly AM (or angle AOM), and he gives a list of values of Om for various values of the true anomaly, at intervals of 10°. The following

[1] Kidd. Hach., xv. 6.

Table,[1] which gives a few of those values compared with the values as found by calculation (for OEC = 30°), shows that whilst agreement is perfect up to 80°, there is a good deal of discrepancy in the case of anomalies 90°, 120°, 150°, 170°. The figures in brackets are those given in a manuscript of the year 1184, which is the earliest MS. of the work in existence. All printed editions give the unbracketted figures.

True Anomaly מסלול הנכון	Equation of the Centre מנת המסלול		
	Calculation	Maimonides	
10°	0°49′	0°50′	
20°	1°37′	1°38′	
30°	2°23′	2°24′	
40°	3° 5′	3° 6′	
50°	3°43′	3°44′	
60°	4°15′	4°16′	
70°	4°41′	4°41′	
80°	4°58′	5° 0′	
90°	5° 8′	5° 5′	
120°	4°39′	4°20′	(4°40′)
150°	2°47′	3°48′	(2°48′)
170°	0°59′	1°50′	(0°59′)

The method of converting mean into true longitude is called **Correction of the Moon** (תיקון הלבנה).

Note: As the maximum equations of the centre for the moon and sun are 5°. 8′ and 1°59′ respectively, it follows that the maximum difference in longitude between a true and a mean conjunction is 7°7′, which at the rate of increase of elongation of 30′28·6″ per hour represents a maximum interval between a true and a mean conjunction of 14 hours. According to Israeli[2], Ptolemy (who gives the maximum equation of the centre for the moon and sun as 5°1′ & 2°23′ respectively) makes the interval equal to 14 hrs. 648 ch. = 14h. 36m.

(B) **Latitude.** Consider Fig. 47, in which ML is the moon's required latitude.

L♈ = Moon's known longitude = 48°35′39″ (p. 145,

$$N\Upsilon = \text{Nodal longitude on same date} = \underbrace{179°2'32''}_{\text{nodal longitude at epoch}}$$

$- 29 \times 3'10{\cdot}63''$ (p. 135) $= 177°30'24''$

$$\therefore \quad LN = L\Upsilon - N\Upsilon = -128°53'45''$$

[1] Kidd. Hach. xv. 6.   [2] Yesod Olam, iii, 14.

Add 360° to make the arc positive and we get $LN = 231°6'15''$. Hence in the spherical triangle MLN, right angled at L, we have :-

Angle at $N = 5°$, and side $LN = 231°6'15'' = (180° + 51°6'15'')$.

And as angle N is so small, MN is practically the same as LN,

$\therefore$ By formula, vi. p. 83,

$$\text{Sin } ML = \text{Sin } 5° \text{ Sin } 231°6'15'' = -\text{ Sin } 5° \text{ Sin } 51°6'15''$$
$$= -0{\cdot}0872 \times 0{\cdot}7783 = -0{\cdot}06787$$

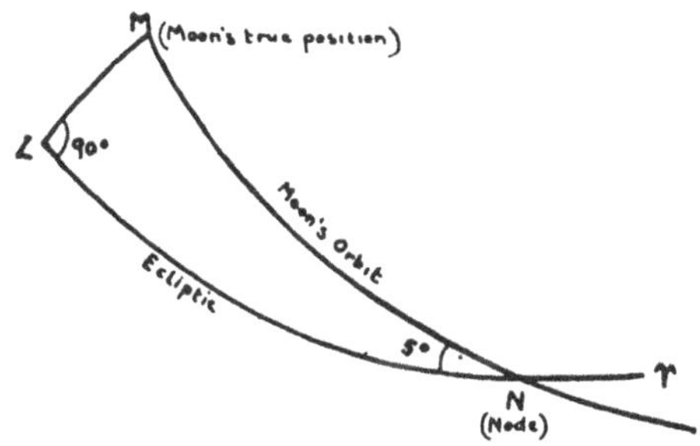

Fig. 47

$\therefore$ $ML = -3°53'29''$, which agrees to the nearest minute with the value $-3°53'$ given by Maimonides.[1] The arc LN (i.e., the difference between the longitude of the moon and that of the node) is called the **Argument of the latitude** (מסלול הרוחב)[2], because it enables one to find the value of the latitude, and Maimonides gives a table showing the values of the lunar latitudes for different values of " Arguments " at intervals of 10°. These are in agreement with calculation to the nearest minute.[3]

[NOTE : Instead of subtracting $29 \times 3'10{\cdot}63''$ from the nodal longitude at the epoch to find N♈, Maimonides[4] adds $29 \times 3'10{\cdot}63''$ to (360° − nodal longitude at epoch), and subtracts the result from 360°. Thus : $\underbrace{360° - 179°2'32''}_{180°57'28''} + 29 \times 3'10{\cdot}63'' = 182°29'36''$;

$360° - 182°29'36'' = 177°30'24'' = N♈$. In this way he avoids working with negative arcs.]

We thus have :-

True Position of Moon on 2nd Iyar, 1178, was { *Longitude* 48°36. *Latitude* − 3°53.

[1] Kidd. Hach., xvi. 19. [2] Ibid, xvi. 10. [3] Ibid xvi. 12. [4] Ibid, xvi. 11.

Strictly speaking, the quantity we have thus found as true longitude of the moon, is not longitude at all, since it is measured along the deferent (which is inclined to the ecliptic) instead of on the ecliptic itself. As however the angle of inclination is so small — never exceeding about 5° — the lengths of these two arcs are practically identical, and hence the arc of the deferent so found may be considered as being actually the true longitude.

It is of course easy to find the actual longitude by solving the triangle NML, but the resulting difference being very small, amounting only to a couple of minutes, Maimonides does not consider it necessary to do this conversion.

To find the Moment of a True Conjunction (i.e. when True Elongation = 0)

In the case we have just considered, the *moon's true longitude* at 34m. 20s. past 6 p.m. on Friday, 2nd Iyar, was 48°35′39″.

We must now find the sun's true longitude at that moment:

(1) Sun's mean longitude (p. 140) 35°38′33″

(2) Longitude of sun's apogee = 86°45′8″ (p. 117) + 29 × 0·15″ = 86°45′12″

∴ Sun's mean anomaly = (1) − (2) = − 51°6′39″

∴ Equation of Centre (see p. 117) = 1·92° Sin (− 51°6′39″) = − 1·92° × ·7784 = − 1°29′40″

∴ *True longitude of sun* = 35°38′33″ − (− 1°29′40″) = 37° 8′13″.

[Maimonides taking the equation of the centre as 1°30′20″,[1] makes the true longitude = 37°9′ (to the nearest minute).[2]]

∴ *True Elongation* = 48°35′39″ − 37°8′13″ = 11°27′26″

At a mean rate of increase of 30′28·6″ per hour (p. 122), the elongation should have taken between 22 and 23 hours to have increased from 0° to 11°27′26″. Let us therefore compute the true longitude of the sun and moon at say 22½ hours before sunset of 2nd Iyar (i.e., at 4m. 20s. past 8 on Thursday, 1st Iyar) as follows:-

A. (1) Sun's mean longitude = 35°38′33″ − 22·5 × 2′28″[3] = 34°43′ 3″

(2) Longitude of Sun's Apogee = 86°45′12″

∴ Sun's mean anomaly = − 52° 2′ 9″

∴ Equation of Centre = − 1·92° Sin 52°2′9″ = − 1·92° × ·7884 = − 1°30′49″

[1] Kidd. Hach., xiii. 4. [2] Ibid, xvii. 14. [3] See p. 113.

∴ *True longitude of sun* $=34°43'3''+1°30'49''=36°13'52''$

B. Moon's mean longitude $=53°36'39''-22{\cdot}5\times32'56''=41°15'39''$

Sun's mean longitude $=34°43'\ 3''$

∴ Double elongation $=2(41°15'39''-34°43'3'')=13°\ 5'12''$

∴ Anomaly excess (Table, p. 144) $=2°$.

Further, mean lunar Anomaly $=103°21'46''$ (p. 143)

$-\ 22{\cdot}5\times32'40''=91°\ 6'40''$

Add Anomaly excess and get, *True Anomaly* $=93°\ 6'40''$

∴ Equation of Centre (Table, p. 146) is about $5°\ 8'$

∴ *True longitude of moon* $=41°15'39''-5°8'=36°10'39''$

∴ True Elongation $=36°10'39''-36°13'52''=-3'13''=-193''$

Hence the true conjunction must have occurred a few minutes *later*, during which small period the rate of increase of true and mean elongation may be taken as sensibly the same, viz., 30·5′ per hour or 30·5″ per minute.

∴ *True Conjunction occurred* $\frac{193''}{0{\cdot}5''}=$ 6m. 20s. later than 4m. 20s. past 8, viz., *at* 10*m*. 40*s*. *past* 8 *on Thursday evening*, or expressed in Jewish time at 2h. 192ch. on Friday (since Friday begins at 6 p.m. on Thursday).

NOTE. From this we learn two things:-

(1) As the mean lunar and solar longitudes at sunset on Friday, 2nd Iyar, were $53°36'39''$ and $35°38'33''$ respectively, therefore the mean elongation was at that time $17°58'6''$ (i.e., $6°30'40''$ greater than the true elongation, as found above) and the mean conjunction (or Molad) occurred $\frac{17°58'6''}{30{\cdot}5'}=\frac{10781'}{305'}$ i.e. 35h. 20m. 51s. before sunset of Friday, viz., 13m. 29s. past 7 a.m. of Thursday. In other words the mean conjunction occurred 12h. 48m. 32s. before the true conjunction.

(2) Had the beginning of that month been calculated from the time of a true conjunction, instead of that of a Molad, Friday would have been the first instead of the 2nd of Iyar.

The moment of a true opposition (i.e., elongation $=180°$) can be computed in a similar manner, and thus the question of the occurrence of an eclipse in that month can be decided (see p. 126). But this point being of no interest in rabbinical astronomy, the calculation is left as an exercise for the reader.

Correction for Parallax. We saw on p. 64 that parallax makes a heavenly body appear lower than it actually is. Parallax, however, has an effect not only upon altitude but also upon

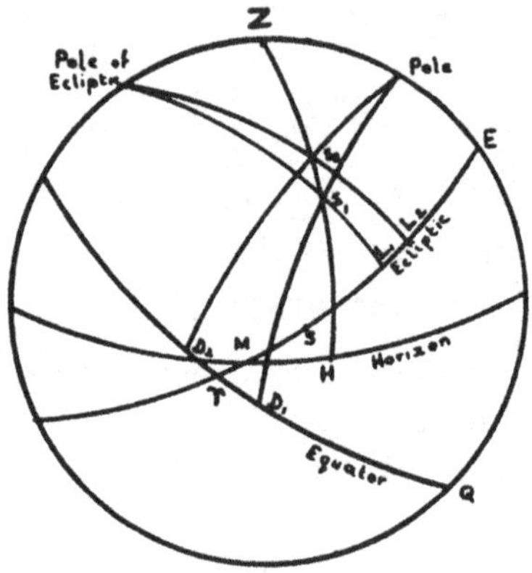

Fig. 48

longitude and latitude, as well as upon right ascension and declination. Thus in Fig. 48, if S_1 and S_2 are the two positions of a body as seen from the surface and the centre of the earth respectively, then the longitudes and latitudes of S_1 and S_2 are ♈L_1 and SL_1 and ♈L_2 and SL_2 respectively. Similarly the right ascensions and declinations of S_1 and S_2 are ♈D_1, SD_1, and ♈D_2, SD_2 respectively. By easy, though somewhat complicated, algebra, it is possible to find formulae for the differences in each case.

In the case of the moon, the differences between the true and apparent longitudes, and between true and apparent latitudes, are given by the formulae:

$$(1) \qquad l - \lambda = P \frac{\text{Cos B}}{\text{Cos b}} \text{Sin} (L - l)$$

$$(2) \qquad b - \beta = P \frac{(l - \lambda) \tan B}{\text{Sin} (L - l)}$$

where l = true longitude, λ = apparent longitude,

b = true latitude, = apparent latitude,

L = longitude of zenith (= ♈E), B = latitude of zenith, and P = horizontal parallax—which may be taken as about 1°.

When the moon is on the horizon it is practically at **M**, the point of intersection of the ecliptic with the horizon, and therefore $l = ♈M$ so that $L - l = ♈E - ♈M = ME = 90°$. We therefore have $\text{Sin}(L - l) = 1$, and the formulae reduce themselves to:-

$$l - \lambda = \frac{\text{Cos B}}{\text{Cos b}}, \text{ and } b - \beta = (l - \lambda) \tan B.$$

Further, as *b* is not greater than about 5°, Cos b is not less than 0·9962 and may therefore be taken as 1. Therefore, the formulae reduce themselves to

$$l - \lambda = \text{Cos B}$$

$$b - \beta = (l - \lambda) \tan B = \text{Cos B} \tan B = \text{Sin B}$$

But B, the latitude of the zenith, $= 90° - \theta$, where θ is the angle of inclination of the ecliptic with the horizon—whose value at any moment can be found from the lunar longitude, l, at that moment (see p. 164).

Hence we have finally:

$$l - \lambda = \text{Cos}(90° - \theta) = \text{Sin}\ \theta$$

$$b - \beta = \text{Sin}(90° - \theta) = \text{Cos}\ \theta.$$

Thus we find from the table on p. 164 that when $l = 0, \theta = 81°30'$.

$$\therefore\ l - \lambda = \text{Sin}\ 81°30' = \cdot9890° = 59'20''$$

$$\text{and } b - \beta = \text{Cos}\ 81°30' = \cdot1478° = 8'52''$$

The following table compares the values of $l - \lambda$, and $b - \beta$ as found mathematically, with those given by Maimonides[1] (the figures in brackets are those in the Oxford MS. of 1184).

l or (360–l)		θ	l – λ			b – β		
			Calculated (= Sin θ)	Maimonides		Calculated (= Cos θ)	Maimonides	
(l)	(360-l)			(l)	(360-l)		(l)	(360″-l)
0		81°30′	59′20″	0°59′		8′52″	9′	
30°	330°	77°54′	58′40″	1° 0′	0°58′	12′35″	10′	12′
60°	300°	67°53′	55′35″	0°58′	0°53′	22′35″	16′	24′ (27′)
90°	270°	54°42′	48′58″	0°43′	0°44′	34′40″	27′	36′
120°	240°	43°21′	41′16″	(52′) 0°43′	0°36′	43′34″	38′	44′
150°	210°	36°38′	35′50″	0°37′	0°34′	48′ 8″	44′	45′
180°		34°30′	33′59″	0°34′		49′27″	46′	

[1] Kidd. Hach., xvii. 5 and 8.

From the table it is seen that the values given by Maimonides are erroneous in two respects. For not only do they differ considerably from the values found by calculation, but—what is more striking—his values, instead of being (as they should be) the same for corresponding values of l and 360° − l, are different.

The second difficulty can, to a certain extent, be overcome if we assume that Maimonides calculated parallax not for the moment of sunset, but for 20 minutes later—which, as he himself states,[1] is the time for which he makes his other calculations,

לאחר שקיעת החמה בכמו שליש שעה

but the figures so obtained differ so widely from those given by Maimonides as to make it improbable that such an explanation is the true one. Indeed it is impossible to account for Maimonides' figures. (See Baneth, Maimuni's Neumondsberechnung, Berlin 1903, p. 148.)

[1] Kidd. Hach. xiv. 6.

CHAPTER XIV.

Kepler's Laws and Gravitation.

With wisdom and with wondrous skill	טובים מאורות שברא אלהינו.
The skies with bright orbs God did fill.	יצרם בדעת בבינה ובהשכל.
He poured therein twain mighty forces,	כח וגבורה נתן בהם
Prepotent o'er their cosmic courses.	להיות מושלים בקרב תבל:
They scatter light—their glow's unfurled,	מלאים זיו ומפיקים נוגה,
Within, without, throughout the world;	נאה זיום בכל העולם;
Obedient to their inner urge, With joy they set, with joy emerge.	שמחים בצאתם וששים בבואם, עושים באימה רצון קונם.
. . . .	. . . .
. . . .	. . . .
God called! the Sun shot forth his rays;	קרא לשמש ויזרח אור
His glance produced the lunar phase.	ראה והתקין צורת הלבנה.

(Sabbath Liturgical Hymn - probably 8th Century).[1]

Whilst a combination of an eccentric with an epicycle suffices to adjust some of the more important irregularities of the moon, the device—however ingenious from a mathematical standpoint—is a purely fantastic one, and has no physical basis whatever. Further, as has already been pointed out (p. 138), on such a theory, the moon's apparent diameter when nearest to, and farthest from, the earth should be in the proportion of about 2 : 1 instead of about 1·14 : 1, which is the actual proportion. Moreover, complicated as we have seen the epicycle theory to be, it becomes still more involved if one tries to explain a few of the numerous irregularities to which the lunar orbit is subjected, for it is then found necessary to postulate the existence of more than one epicycle.

In the beginning of the 17th century, Kepler—after a most patient and careful study of the motion of the planets—swept away, once and for all, the whole complicated theory of epicycles, and showed conclusively that each of the planets moved in the

[1] Translated by Reginald V. Feldman, B.A.

form of an ellipse round the sun as a common focus. The moon—which is a satellite of the earth — has the earth at one of the foci of its elliptical orbit. The shape of the ellipse, however, is different in each case, the differences lying in the sizes and directions of their major axes, the values of their eccentricities, and the angles of their inclinations with the ecliptic, as the following table shows:-

	Earth	Moon	Mercury	Venus	Mars	Jupiter	Saturn
Major Axis	10	—	3.87	7.23	15.23	52	95.3
Eccentricity	0.0168	0.055	0.205	0.007	0.093	0.048	0.056
Inclination to Ecliptic	0	5°	7°'6"	3°23'28"	1°51'6"	1°18'51"	2°29'36"

Kepler's Laws of Planetary Motion. are:

(1) Every planet describes an ellipse round the sun (S) as one of its foci. This law follows from the observed fact that the actual rate of planetary motion at any moment is inversely proportional to the square of its distance from the sun at that moment. (See note on p. 110.)

(2) The straight line joining the centre of the sun (S) to that of the planet sweeps out equal areas in equal times. Thus, if A_1B_1, A_2B_2, and A_3B_3 (Fig. 49) be the arcs of the ellipse described by the planet in equal times at different parts of the orbit between aphelion (C) and perihelion (D), then the areas SA_1B_1, SA_2B_2, and SA_3B_3 are equal to one another. This can be shown to be a direct consequence of the first law.

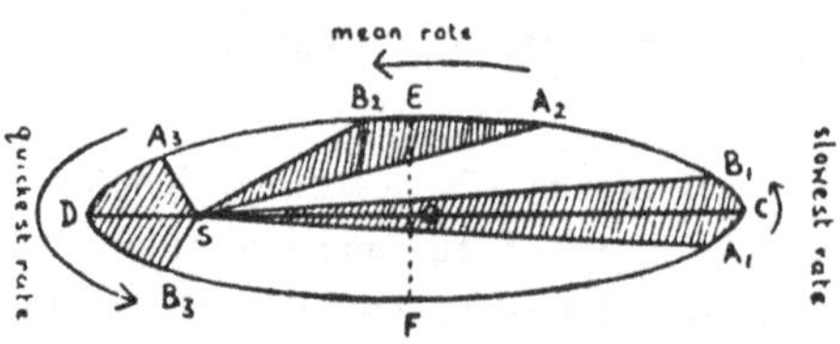

Fig. 49

(3) The squares of the times of revolution of any two planets about the sun bear to the cubes of their respective mean distances

from the sun, a constant ratio. $\left(\frac{T_1^2}{d_1^3} = \frac{T_3^4}{d_2^3} = \; ..... = K\right)$

Corollary. From this it follows that the time of revolution of any planet about the sun, i.e., its periodic time T, is ascertainable from its mean distance d from the sun.

Thus, if we take the mean distance of the earth from the sun as 10 and that of Mercury as 3·87; the periodic time of the earth as 365·25 days and that of Mercury as 87·97 days, we get—

$$\frac{(365{\cdot}25)^2}{10^3} = \frac{133410}{1000} = 133:1 \text{ (in the case of the earth)}$$

$$\text{and } \frac{(87{\cdot}97)^2}{(3{\cdot}87)^3} = \frac{7738{\cdot}7}{58{\cdot}27} = 133:1 \text{ (in the case of Mercury)}$$

Hence, if, for instance, the mean distance of Venus from the sun is known to be 7·23, we would expect its periodic time T to be such that $\frac{T^2}{(7{\cdot}23)^3} = 133$, whence $T = 224{\cdot}2$ days. (As a matter of fact, observation shows the exact periodic time of Venus to be 224·69 days.)

Per Contra, if the periodic time of a planet is known, its mean distance from the earth can be found.

This theory of elliptical motion is not only simple, but has the merit of being true; for although Kepler merely announced his laws without being able to explain them, Newton—towards the end of the 17th century — showed that all these laws are the necessary consequences of his own universal law of gravitation — which is: that the force of attraction between two bodies of masses m_1 and m_2 separated by a distance d is proportional to $\frac{m_1 m_2}{d^2}$. Further, with elliptical orbits, the numerous irregularities in the path of the moon (called **perturbations**) are found to be just what one would *a priori* expect as logical results of Newton's law. This is not the place to enter into a full discussion of the *problem of three bodies*, which analyses the effect, upon one body, of the combined gravitational attraction of two other bodies, pulling upon it with different forces in different directions, but we shall, briefly, consider a few of the more important consequences of the mutual attraction of the sun, earth and planets upon the moon :-

(1) **Variation.** When the sun, moon and earth, or sun, earth and moon, are in the same straight line — as they are in conjunction or opposition respectively, one of the two (moon or earth) is nearer the sun than the other, and is consequently attracted to it more, with the result that the distance between the earth and the moon becomes increased and the attraction between them becomes diminished. Again, when the moon is in quadrature, it is at the same distance from the sun as is the earth, and both being pulled with the same force towards the sun along two convergent lines, the distance between the moon and the earth becomes in consequence diminished and the attraction between them is increased. Hence, certain disturbances in the moon's path must occur as the result of the change of angular distance (or elongation E), between the sun and the moon. This alteration, or perturbation, is called *Variation* (V), and can be shown to be represented by the equation $V = 35{\cdot}7' \text{ Sin } 2E$, where V represents the difference between the moon's mean and true positions. From this it follows that in syzygy (i.e., conjunction or opposition), when $\text{Sin } 2E = 0$, V is zero, i.e., in those positions, the moon's true position coincides with its mean position. Similarly, in quadrature (where $E = 90°$ or $270°$), $\text{Sin } 2E = 0$, and V is again zero. Its maximum positive or negative effects occur when the moon is midway between conjunction, or opposition, and a quadrature, i.e., at the octants, when $E = 45°$, or $135°$, etc. Then $\text{Sin } 2E = \pm 1$, and therefore $V = \pm 35{\cdot}7'$.

Variation was discovered in the 16th Century by Tycho Brahé, who studied the moon in all parts of its course, although it was possibly known to the Arabian astronomer Abu'l Wafa 500 years earlier. The early Greek astronomers, however, who observed the moon only during syzygy or quadrature, missed this perturbation—because in those positions, as we have seen, there is no difference to be detected between the moon's true and mean positions.

(2) **Evection.** As the moon's distance from the earth is greatest in apogee, and least in perigee, the attraction between them is least in the former, and greatest in the latter position, and will be of intermediate magnitudes between apogee and perigee. In other words the attraction between moon and earth will depend upon the moon's anomaly (or its angular distance

from its apogee), with resulting disturbances in the lunar orbit. This difference between the moon's true and mean positions depending upon the eccentricity of its orbit is called *evection* (e), and, as we have just seen, depends in some way upon the lunar anomaly A. It also depends upon its elongation E, and can be shown to be represented by the equation e = 1·3° Sin (2E–A). At syzygy, therefore, when 2E = 0, e = –1·3° Sin A. It is this perturbation that Ptolemy corrected to some extent by means of the epicycle and prosneusis—although it was previously detected by Hipparchus, who had found discrepancies between the times of calculated and observed eclipses. If 2E – A is less than 180°, then evection has to be subtracted; if it is greater than 180°, evection has to be added.

(3) **Annual Equation.** As the distance of the earth from the sun is also variable, depending upon the solar anomaly, or angular distance between the sun and its apogee, it follows that the sun's pull upon the earth, and consequently the earth's pull upon the moon, must vary—throughout the year—with the sun's anomaly. There are, accordingly, disturbances in the moon's orbit varying with the solar anomaly (a). This kind of disturbance is called *Annual Equation* (A.E.) and is represented by the formula A.E. = 11·15′ Sin a. When Sin a = ±1, A.E. has its maximum positive or negative value of 11·15′ or 11′9″.

(4) **Secular Acceleration.** The planets pull upon the sun drawing it away from the earth, thus diminishing the pull between the earth and the moon (i.e., diminishing the eccentricity of the earth's orbit), and causing an *acceleration* in the latter's mean motion. This perturbation, which is very slight, amounting to an advance of about 8″ in 100 years in the moon's mean position, tends to produce a progressive shortening of the month. The phenomenon was discovered by Halley, who, in comparing ancient with modern eclipses, found the month to be gradually shortening. As the quickened motion corresponds to a shortening of the lunar orbit, it follows that if this acceleration were to continue indefinitely, the moon would ultimately reach the earth. It was shown by Laplace, however, that after an interval of some 25,000 years from now, when the eccentricity of the *lunar* orbit will reach its minimum of 0·003, the acceleration will change into a retardation and the moon will begin to recede again from the

earth for 40,000 years, by which time the eccentricity will reach its maximum of 0·02, when shortening of the month will begin again. There are many other perturbations (more than 100) to which the moon's orbit is subjected, but they are too small to be considered here.

The **Equation of the Centre of the Moon** (for an elliptical orbit) is given by $114{\cdot}6° \text{ e Sin A} + 71{\cdot}63° \text{ e}^2 \text{ Sin 2A}$, which—as $e = 0{\cdot}055$ — becomes $6{\cdot}3° \text{ Sin A} + 0{\cdot}2166° \text{Sin 2A}$ (where A = mean lunar anomaly).

Calculation of Solar and Lunar Longitude on basis of an Elliptical Orbit.

We shall now calculate the true longitude of the sun and moon on 2nd Iyar, 1178, on the theory of an elliptical orbit, and compare the results with those found by means of the epicycle method :

	Mean Longitude of Sun	Mean Solar Apogee	Mean Longitude of Moon	Mean Lunar Anomaly
Epoch	7° 3′32″	86°45′ 8″	31°14′43″	84°28′42″
29 days	28°35′ 1″	4″	22° 6′55″	18°53′ 4″
2nd Iyar	35°38′33″	86°45′12″	53°21′38″	103°21′46″

Hence, Mean Solar Anomaly = 35°38′33″ − 86°45′12″ = − 51°6′39″

∴ Equation of Centre (for Sun) = − 1·92 Sin 51°6′39″
= − 1·92 × ·7784 = − 1°29′40″

∴ **True Longitude of Sun** = 35°38′33″ + 1°29′40″
= 37°8′13″ (see p. 148)

Mean Elongation = 53°21′38″ − 35°38′33″ = 17°43′5″ = E

∴ *Variation* (= 35·7′ Sin 2E) = 35·7′ Sin 35°26′10″ = 35·7′ × ·5798
= 20′42″

Also, *Annual Equation* = − 11·15 Sin 51°6′39″ = − 11·15′ × ·7784
= − 8′41″

And, *Evection* = 1·3°Sin(2E − A) = 1·3°Sin(35°26′10″ − 103°21′46″)
= − 1·3 Sin 67°55′36″ = − 1·3° × ·9267 = − 1°12′17″

∴ **Corrected lunar longitude**

$$= \underbrace{53°21'38''}_{\text{Mean Longitude}} + \underbrace{20'42''}_{\text{Variation}} - \underbrace{8'41''}_{\text{A.E}} + \underbrace{1°12'57''}_{\text{Evection}}$$

$$= \mathbf{54°45'56''}$$

Also, **Corrected lunar Anomaly**

$$= \underbrace{103°21'46''}_{\text{Mean Anomaly}} + \underbrace{20'42''}_{\text{Variation}} - \underbrace{8'41''}_{\text{A.E}} + \underbrace{1°12'17''}_{\text{Evection}}$$

$$= \mathbf{104°46'4''}$$

∴ Equation of Centre (for Moon)

$$= 6{\cdot}3° \text{Sin } 104°46'4'' + 0{\cdot}2166 \text{ Sin } 209°32'8''$$
$$= 6{\cdot}3° \text{Sin } 75°13'56'' - 0{\cdot}2166 \text{ Sin } 29°32'8''$$
$$= 6{\cdot}3° \times {\cdot}9670 - 0{\cdot}2166° \times {\cdot}4930$$
$$= 5°59'7''$$

∴ **True lunar longitude**$= 54°45'56'' - 5°59'7'' = 48°46'49''$ which differs from $48°35'39''$ found by the epicycle method (p.145), by no more than about 11 minutes.

CHAPTER XV.

The Visibility of the Moon.

כך אמרו חכמים, הראה לו הקב"ה למשה במראה הנבואה דמות לבנה ואמר לו כזה ראה וקדש (קדוש החדש, א.א)

"The sages have said (see Menachoth 29a): God showed to Moses in a prophetic vision, a figure of a moon's crescent and said to him, when thou seest the new moon like this, sanctify it," (Kidd. Hach. i. 1)

Having found the moon's longitude and latitude, the Talmudic Rabbis next had to ascertain (see pp. 5 and 112) whether in that position in the heavens the new moon would be visible in the neighbourhood of Jerusalem. For, as the beginning of a month was fixed on the accredited evidence of witnesses who reported having seen the new moon soon after sunset on a certain day, it was the duty of the Calendar Council not only to test their evidence by stringent cross-examination (see next chapter), but also to ascertain, by mathematical calculation, whether the moon could, in fact, be seen at that particular moment at the particular place from which the witnesses came. The following additional data were required to answer this question:-

(1) **The True Elongation,** (i.e., the difference in true longitudes of the moon and sun), which Maimonides calls the first length[1] אורך ראשון

(2) **The Arc of Vision,** i.e., the difference in altitudes between the sun and the moon.

"If," says Maimonides,[2] "the true elongation is no more than 9°, then further calculation is unnecessary, since, with such an elongation the lunar crescent is so small that it is quite impossible for it to be visible anywhere in Palestine (see p. 177). If the elongation is greater than 15°, it is equally unnecessary to make any further calculation, since one may be quite sure that the crescent would be so large that—under suitable meteorological conditions—it would not fail to be visible throughout Palestine. It is only for intermediate values of the elongation that further

[1] Kidd. Hach., xvii. 1. [2] Ibid xvii. 3 & 4.

mathematical investigation is necessary. This, however, is only the case if the moon's true longitude lies between 270° (1st of Capricorn) and 90° (1st of Cancer). For longitudes between 90° and 270° (Cancer to Capricorn), the corrresponding true elongation limits are 10° and 24° respectively." (For an explanation of the influence of lunar longitude on elongation limit, see p. 165)

For a true elongation between these limits (9° to 15°, or 10° to 24°, according to the size of the moon's longitude), the next step is to ascertain whether the difference in altitude between the sun and moon is sufficiently large for the sun not to outshine the moon. For, in the same way as there must be a certain depth of the sun below the horizon to produce enough darkness to render the stars visible (see p. 93), so also must there be a sufficient vertical distance between the moon and the sun for the former to be seen—provided its crescent, whose size depends upon the elongation (see p. 176) is sufficiently large. As the Talmud puts it (in an allegorical discussion between the moon and God):[1] "Of what use is a candle light on a bright sunny day?" (שרגא בטיהרא מאי אהני). This vertical luni-solar distance is what should rightly be called the *Arc of Vision* (קשת הראיה), but Maimonides uses this term for a certain arc of the equator from which this vertical distance can be calculated (see p. 170).

It is obvious that the greater the lunar crescent, the less need be the vertical luni-solar distance; and hence, the greater the elongation — between the above-mentioned limits — the less the necessary vertical luni-solar distance. If, however, the elongation is less than 9° or 10° (according to the lunar longitude), the crescent will not be visible, however large the vertical luni-solar distance.

Calculation of the Arc of Vision.

In Fig. 50, S and L are the positions of the sun and moon respectively.

(1) LB = lunar latitude (b)

(2) ♈B = lunar longitude (l)

(3) ♈S = solar longitude (λ)

and (4) SB = elongation ($l - \lambda$)

[1] Chulin, 60 b.

To simplify matters, we shall assume the point B to be on the horizon.

SP = depth of sun below horizon (a), so that angle SPB = 90°
LH = altitude of moon above horizon (h), so that angle LHB = 90°

∴ Vertical Luni-Solar Distance, or Arc of Vision = LH + PS

(i) **Evaluation of LH.** Assume that the angle of inclination (θ) between the ecliptic and the horizon is known. Then, in the spherical triangle LHB (right-angled at H) we have:-

Angle LBH = $(90° - \theta)$, and LB is known;

∴ By formula vi., p. 83, Sin LH = Sin LB Sin $(90 - \theta)°$
= Sin LB Cos θ

∴ LH is determined (if we know θ).

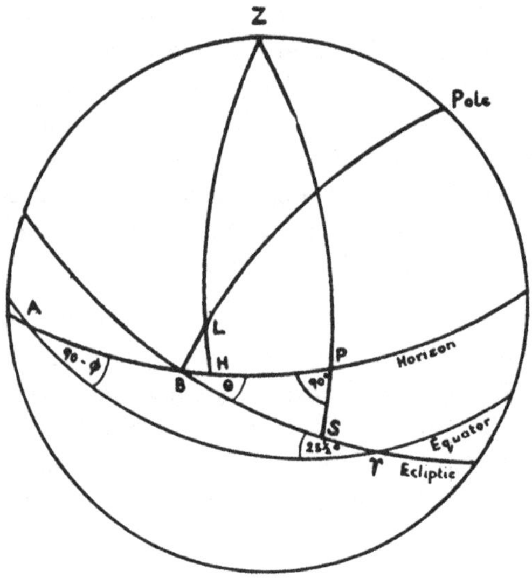

Fig. 50

(ii) **Evaluation of PS.** Similarly in triangle SBP, we have:-
Angle SPB = 90°; angle SBP = θ, and side SB = known elongation.

∴ By the same formula, Sin PS = Sin SB Sin θ, whence PS is determined.

Therefore, the whole Arc of Vision LH + PS is determined, if we know the value of θ at the moment in question. This angle has no fixed value at any given place but varies from moment to moment. Its value, however, depends upon the geographical latitude (ϕ) of the place as well as upon the longitude of the moon at the moment. To evaluate it we may proceed as follows:-

Calculation of θ. In triangle ♈AB, we have
Angle at A $=(90°-\phi)$; angle at ♈ $=23\frac{1}{2}°$, and side ♈B = known lunar longitude.

$\therefore$ From the well-known property of any spherical triangle ABC, viz., Cos A = Sin B Sin C Cos a − Cos B Cos C, we have, (in Fig. 50),

Cos A = Sin B Sin $23\frac{1}{2}°$Cos ♈B − Cos B Cos $23\frac{1}{2}°$ (where B $=180°-\theta$)

$\therefore$ In the case of Jerusalem (where $\phi=32°$ and A $=90°-32°=58°$),

Cos 58° = Sin $(180°-\theta)$ Sin $23\frac{1}{2}°$Cos ♈B − Cos $(180°-\theta)$ Cos $23\frac{1}{2}°$

Dividing throughout by Cos $23\frac{1}{2}°$, we get

$$\frac{\text{Cos } 58°}{\text{Cos } 23\frac{1}{2}°}\left(=\frac{\cdot 5289}{\cdot 9171}=0\cdot 5778\right)$$

= Sin $(180°-\theta)$ tan $23\frac{1}{2}°$Cos ♈B − Cos $(180°-\theta)$

i.e., 0·5778 = ·4348 Cos ♈B Sin $(180°-\theta)$ − Cos $(180°-\theta)$.

As ♈B, and therefore Cos ♈B, is known, Sin $(180°-\theta)$, and therefore θ is ascertained.

The best method of solving this equation is to look up in trigonometrical tables an angle whose Cotangent is equal to 0·4348 Cos ♈B. Let this angle be α; the equation then becomes

0·5778 = Cot α Sin $(180°-\theta)$ − Cos $(180°-\theta)$.

Multiply throughout by Sin α and get:

0·5778 Sin α = Cos α Sin $(180°-\theta)$ − Cos $(180°-\theta)$ Sin α
= Sin $(180°-\theta-\alpha)$

Thus, in our case, where ♈B = 48°36′ and therefore Cos ♈B = 0·6613, we have,

0·4348 Cos ♈B = 0·4348 × ·6613 = 0·2843

$\therefore$ Cot α = 0·2843, whence α = 74°8′

$\therefore$ 0·5778 Sin α = 0·5778 Sin 74°8′ = 0·5778 × ·9619 = 0·5558

$\therefore$ Sin $(180°-\theta-\alpha)$ = 0·5558, whence $180°-\theta-\alpha$ = 33°55′

$\therefore$ $180°-\theta-74°8′$, i.e., $105°52′-\theta$ = 33°55′

$\therefore$ θ = 105°52′ − 33°55′ = **71°57′**

Having found θ, we now evaluate LH, and PS.

(i) **Sin LH = Sin LB Cos θ = Sin (– 3°53′) Cos 71°57′**

= – 0·0678 × 0·3098 = – 0·0210

$\therefore$ **LH = – 1°12′**

(ii) **Sin PS = Sin SB Sin θ = Sin 11°27′ Sin 72°6′**

= 0·1986 × 0·9516 = 0·1890

$\therefore$ **PS = 10°54′**

$\therefore$ Arc of Vision = LH + PS = – 1°12′ + 10°54′ = 9°42′.

Subtracting 1° for horizontal parallax, we get :- **Corrected Arc of Vision = 8°42′.**

From the equation :

0·5778 = 0·4348 Cos♈B Sin (180° – θ) – Cos (180° – θ) (p. 163)
we see that the value of θ at any moment depends upon the longitude of the moon ♈B at that moment. As, further, the values of LH and PS depend upon that of θ, it is clear that θ, LH, and PS must vary with lunar longitude. Substituting various values of ♈B in the above equation, we get the following Table, which tells us, that when ♈B lies between 270° and 90°, $\theta > 58°$ for the latitude of Jerusalem, whilst if ♈B lies between 90° and 270°, $\theta < 58°$.

l (= ♈B)	θ	a (= PS)	h (= LH)
0° or 360°	81°30′	0·9886 SB	0·1478LB
30° or 330°	77°54′	0·9769 SB	0·2096 LB
60° or 300°	67°53′	0·9236 SB	0·3765 LB
90° or 270°	54°42′	0·8104 SB	0·5779 LB
120° or 240°	43°21′	0·6789 SB	0·7272 LB
150° or 210°	36°38′	0·5888 SB	0·8241 LB

From Fig. 51, as well as from equation Sin PS = Sin SB Sin θ it is clear that for constant PS (solar depth below the horizon), the length of SB (true elongation) varies inversely with the size of the angle θ. Hence it follows that if an elongation SB of 9° is necessary to make PS of the requisite size to render the moon visible when $\theta > 58°$, the elongation SB_1 necessary to make PS of the same requisite size when $\theta_1 < 58°$, will have to be greater than 9°. This explains the statement of Maimonides quoted on

p. 161 to the effect that when lunar longitude lies between Capricorn and Cancer (i.e., $\theta > 58°$), an elongation of not less than 9° is necessary for visibility, whilst for a lunar longitude between Cancer and Capricorn the moon will not be visible unless SB exceeds 10°.

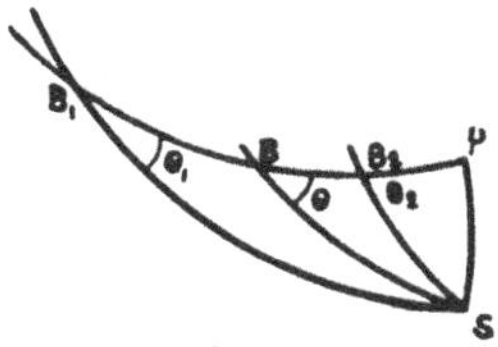

Fig. 51

Note. It will be remembered that when speaking of *mean* lunar longitude (p. 144) Maimonides gave the minimum value of the "double elongation" compatible with visibility as 5°, and the mean "double elongation", beyond which the moon cannot fail to be seen, as 62°, from which it follows that a *mean elongation of 2½° may correspond to a true elongation of 9°, and one of 31°, may correspond to a true elongation of 24°* (see p. 161). Now, in view of the fact that the maximum equation of the centre in the case of the sun is 1°59′ (p. 117), and that in the case of the moon is 5°8′[1] the maximum difference between the respective true and mean elongation limits should be 1°59′ + 5°8′ = 7°7′, i.e. 7° to the nearest ½°. This holds good in the case of the major limits (31° — 24° = 7°), but in the case of the minor limits, there is a discrepancy of at least ½°, since 9° —2°30 = 6°30′, and not 7°.

The explanation of this apparent discrepancy lies in the fact that Maimonides does not give the elongation limits, but the "double elongation" limits — which he gives to the nearest ½° as 5°, and 62°. Therefore 5°, would also include 4°30′. corresponding to an elongation of 2°15′ — which must therefore be taken as the minor elongation limit. Hence the difference between the two minor elongation limits is 9°—2°15′ = 6°45′, which is the same as 7°, to the nearest ½°.

Maimonides' Method, The method just described for computing the luni-solar vertical distance is direct and simple, and does away with the necessity of finding parallax in latitude and longitude; horizontal parallax in altitude, (whose value is about 1°), is the only correction for parallax to be taken into account at the end of the calculation. Maimonides, however, uses a far more complicated method consisting of the following steps (Kidd. Hach. xvii):-

[1] This is the value for a true anomaly of 100° as given in all editions of Kiddush Hachodesh (xv. 6) as well as in most of the manuscripts, and agrees with the calculated value. The Oxford MS. of 1184, however, gives the value as 5°18′ which is almost certainly a *lapsus calami* for 5°8′.

(1) Having found the true lunar longitude ♈B, he deducts the appropriate parallax in longitude (שינוי מראה האורך) to get the apparent longitude ♈B_1 (Fig. 52), which, after deducting the true solar longitude ♈S, gives the apparent elongation SB_1—which he calls the *Second length* (אורך שני).

(2) He then converts true into apparent latitude L_1B_1—by subtracting, or adding, the appropriate parallax in latitude, according as the latitude is positive (north), or negative (south). The apparent latitude thus found, he calls *Second breadth* (רוחב שני).

(3) Maimonides next estimates—without showing the manner of doing so—the size of the ecliptic arc CB_1 (where C is the point of the ecliptic through which passes the moon's parallactic declination arc L_1D_1). By deducting this arc from, or adding it to, SB_1— according as L_1B_1 is positive, North, or negative, South — he gets the value of the ecliptic arc SC, which he calls the *third length* (אורך שלישי).[1]

The following is the method of finding the value of CB_1 :-

(a) In spherical triangle CD_1♈,

$$\text{Cot } C = \text{Cot } ♈C \tan 23\tfrac{1}{2}^\circ = {\cdot}4348 \text{ Cot } ♈C$$

(b) In spherical triangle CB_1L_1,

$$\text{Cot } C = \text{Cot } L_1B_1 \text{ Sin } CB_1 = \frac{\text{Sin } CB_1}{\tan L_1B_1}$$

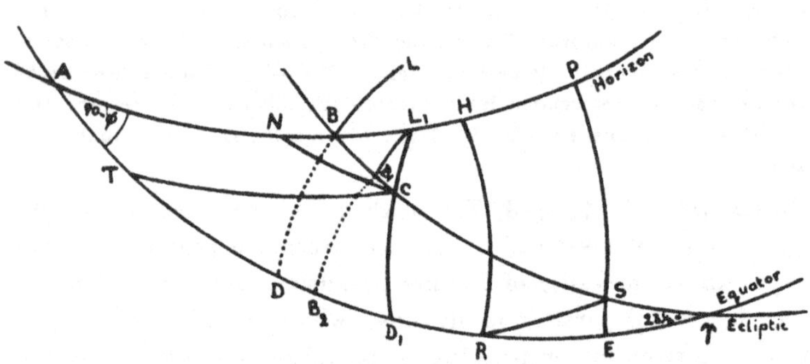

Fig. 52

$$\therefore \quad \frac{\text{Sin } CB_1}{\tan L_1B_1} = {\cdot}4348 \text{ Cos} ♈C$$

This is the case when lunar longitude is between 270° and 90°. When the longitude lies between 90° and 70°, the rule about adding or subtracting is reversed (xvii 11)

But since L_1B_1 is never greater than about 5° (p. 121), therefore CB_1 must be less than 5°, and since the Sines and Tangents of small angles are proportional to the angles themselves,

$$\therefore \quad \frac{\text{Sin } CB_1}{\tan L_1B_1} = \frac{CB_1}{LB_1} = \cdot 4348 \text{ Cos } ♈C$$

$$\therefore \quad CB_1 = \cdot 4348 \; LB_1 \text{ Cos } ♈C$$

But CB_1 being very small, $♈C$ is practically the same as $♈B_1$ (the moon's apparent longitude).

$$\therefore \quad \mathbf{CB_1} = \cdot 4348 \; L_1B_1 \text{ Cos } ♈B_1$$

Maimonides calls CB_1 *the fraction of the apparent latitude.* (מקצת הרוחב השני. or מעגל, or נליזת מעגל הירח).

Example. To find the value of CB_1 when lunar longitude lies between 0° and 20°.

When $♈B_1 = 0$, Cos $♈B_1 = 1$ $\therefore$ $CB_1 = \cdot 4348 \; L_1B_1$

When $♈B_1 = 20°$, Cos $♈B_1 = \cdot 9397$ $\therefore$ $CB_1 = \cdot 4086 L_1B_1$

$$\therefore \quad \textbf{Mean Value of } \mathbf{CB_1} = \frac{(\cdot 4348 + \cdot 4086)}{2} L_1B_1 = \mathbf{\cdot 4217 L_1B_1}$$

The table on p. 168 compares the value of CB_1 (in terms of L_1B_1, or lunar apparent latitude) for different values of lunar longitude, as found mathematically, and as given by Maimonides (xvii, 10).

(4) Maimonides next calculates the value of the arc of the equator $♈T$ corresponding to $♈C$ (where CT is parallel to the horizon). The manner of calculating it—which he does not give—is as follows:-

In the right angled spherical triangle $CD_1♈$ ($D = 90°$), the angle at $♈ = 23\frac{1}{2}°$, and the side $♈C$ has just been found.

$$\therefore \quad \text{Cos } CD_1 = \frac{\text{Cos } ♈C,}{\text{Cos } ♈D,}$$

$$\text{But Sin } CD_1 = \text{Sin } ♈C \text{ Sin } ♈,$$

$$\therefore \quad \text{By division, } \tan CD_1 = \tan ♈C \text{ Sin } ♈ \text{ Cos } ♈D_1.$$

But in right angled triangle CD_1T,

$$\text{Sin } D_1T = \tan CD_1 \tan \phi \text{ (where } \phi \text{ is geographical latitude)}$$

Table giving the values of CB_1 for different values of Lunar Longitude.

Moon's apparent longitude (אורך שני)	Value of CB_1 (מעגל) in terms of L_1B_1	
	Calculated	Given by Maimonides
0°- 20° (180°-200°) or 340°-360° (160°-180°)	$0{\cdot}42\ L_1B_1$	$\frac{2}{5}\ L_1B_1$
20°- 40° (200°-220°) or 320°-340° (140°-160°)	$0{\cdot}37\ L_1B_1$	$\frac{1}{3}\ L_1B_1$
40°- 50° (220°-230°) or 310°-320° (130°-140°)	$0{\cdot}31\ L_1B_1$	$\frac{1}{4}\ L_1B_1$
50°- 60° (230°-240°) or 300°-310° (120°-130°)	$0{\cdot}25\ L_1B_1$	$\frac{1}{5}\ L_1B_1$
60°- 70° (240°-250°) or 290°-300° (110°-120°)	$0{\cdot}18\ L_1B_1$	$\frac{1}{6}\ L_1B_1$
70°- 80° (250°-260°) or 280°-290° (100°-110°)	$0{\cdot}11\ L_1B_1$	$\frac{1}{12}\ L_1B_1$
80°- 85° (260°-265°) or 275°-280° (95°-100°)	$0{\cdot}06\ L_1B_1$	$\frac{1}{24}\ L_1B_1$
85°- 95° (265°-275°) or 265°-275° (85°- 95°)	$0{\cdot}0$	0

$\therefore$ $\text{Sin } D_1T_1 = \tan \text{♈}C \text{ Sin ♈ Cos ♈}D_1 \tan \phi$

Now, $\tan \text{♈}D_1 = \tan \text{♈}C \text{ Cos ♈}$, $\therefore$ $\tan \text{♈}C = \dfrac{\tan \text{♈}D_1}{\text{Cos ♈}}$

$$\therefore\ \textbf{Sin } \mathbf{D_1T} = \frac{\tan \text{♈}D_1}{\text{Cos ♈}} \text{Sin ♈ Cos ♈}D_1 \tan \phi$$

$$= \tan \text{♈}D_1 \tan \text{♈ Cos ♈}D_1 \tan \phi$$

$$= \tan 23\tfrac{1}{2}^\circ \tan 32^\circ \text{Sin ♈}D_1$$

$$= {\cdot}4348 \times {\cdot}6249 \text{ Sin ♈}D_1$$

$$= \mathbf{{\cdot}2717\ Sin\ \text{♈}D_1}$$

Now, since $\tan \text{♈}D_1 = \tan \text{♈}C \text{ Cos } 23\tfrac{1}{2}^\circ$,[1] and ♈C is known, therefore $\text{♈}D_1$ is known, and therefore also $\text{Sin ♈}D_1$ is known. Hence $\text{Sin } D_1T$, and therefore D_1T is known. Therefore $\text{♈}T = \text{♈}D_1 + D_1T$ is found.

[1] By a well known property of a right angled spherical triangle.

NOTE: The arc ♈T is called the moon's *oblique ascension* (מצער עקום) in contra-distinction to the right ascension ♈D_1 (מצער ישר). The arc D_1T is therefore called the *Ascension difference* (הבדל המצערים). Similarly in the case of the sun, if SR is parallel to the horizon, and SE is the sun's declination, then ♈R = sun's oblique ascension; ♈E = right ascension and ER = ascension difference.

Example. What is the moon's oblique ascension ♈T, when ♈C (the 3rd length) = 30°.

Tan ♈D_1 = tan ♈C Cos ($\frac{23^\circ}{2}$, $\frac{30^\circ}{2}$) Cos $23\frac{1}{2}$ = ·5774 × ·9171 = ·5295.

∴ ♈D_1 (Moon's apparent right ascension) = 27°54′.

∴ Sin D_1T = ·2717 Sin 27°54′ = ·2717 × ·4679 = ·1271, whence D_1T = 7°18′.

∴ ♈T = ♈D_1 + D_1T = 27°54′ + 7°18′ = 35°12′,

i.e. When ♈C = 30°, the moons oblique ascension in $\frac{7}{6}$ of ♈C, which is the value given by Maimonides, who tabulates the oblique ascensions of the moon for different values of ♈C.[1] His values are in substantial agreement with those found by calculation as above.

(5) Having found ♈T, we can find ♈A, if we know the value of TA, as follows:-

Draw CN parallel to the equator, cutting the horizon at N.

As the arcs L_1C, CN, and NL are very small, the triangle L_1CN may treated as plane. Also CN being parallel to the equator, the angle $L_1CN = 90^\circ$, and angle $L_1NC = 90^\circ - \phi$, whence angle $NL_1C = \phi = 32^\circ$ (in the case of Jerusalem).

$$\therefore \frac{NC}{CL_1} = \tan 32^\circ = \cdot6249, \text{ and } NC = \cdot6249\ CL_1 = \text{about } \tfrac{2}{3}\ CL_1$$

$$\therefore \text{TA (which} = \text{NC)} = \text{about } \tfrac{2}{3}\ CL_1$$

As the value of TA thus depends upon the geographical latitude of the place, Maimonides calls this arc the *quota of the geographical latitude*[2] (מנת גובה המדינה). Further, as L_1C is very nearly equal to L_1B_1, which in turn is very nearly equal to LB, therefore Maimonides gives the value of TA as $\frac{2}{3}$ of the "first breadth" (or real latitude).

The value of TA has to be added to ♈T, or subtracted from it, according as LB is positive (North) or negative (South), to obtain the length of the arc ♈A.[3]

[1] Kidd. Hach. xvii. 12. [2] Ibid. [3] Ibid.

(6) Maimonides now finds the oblique ascension of the sun ♈R, which is done in a similar manner to that outlined. Thus ♈E, the sun's right ascension is found by solving the right angled triangle ♈ES (since ♈S= sun's longitude, is known, and ♈ $=23\frac{1}{2}°$). Also ER is found by solving the right angled triangle SER, of which SE, the sun's declination, is known (from the solution of triangle ♈ES, and angle SRE $=90°-\phi$ $(=58°)$. Therefore ♈R= YE+ER is known. Subtraction of ♈R from ♈T gives the arc RT, which is the difference in oblique ascensions between moon and sun. This arc RT is called the *fourth length* (אורך רביעי).[1]

(7) Knowing RT and TA, we know the value of RA, which Maimonides calls the **Arc of Vision** (קשת הראיה),[2] because it enables one to calculate the luni-solar vertical distance RH, or SP, upon which depends the visibility of the moon. For in triangle RHA, RA is known; angle $H=90°$, and angle $A=58°$.

$$\therefore \quad \text{Sin RH} = \text{Sin RA Sin } 58° = \cdot 8480 \text{ Sin RA}$$

Maimonides does not go on to work out the value of RH, but gives the limits of RA below which the moon cannot be seen, and above which it cannot fail to be seen. Thus, if RA is less than 9°, the moon will be invisible whatever the elongation; if RA>14°, then, provided the elongation is greater than 9°, the moon will be visible throughout Palestine. Between these limits we get the following table:-[3]

Arc of Vision (קשת הראיה)	9°-10°	10°-11°	11°-12°	12°-13°	13°-14°
Elongation (אורך ראשון)	13° or more	12° or more	11° or more	10° or more	9° or more

Example. Maimonides computes the Arc of Vision RA for sunset of Friday, 2nd Iyar, 1178, as follows.[4] (The adjacent column gives for comparison the corresponding data derived mathematically as described above.):-

[1] Kidd. Hach. xvii. 12. [2] Ibid. [3] Ibid xvii, 15—21.
[4] Kidd. Hach. xvii. 13—14.

		Maimonides	Calculation
A.	True Longitude of Moon (♈B)	48°36′	48°36′
	,, ,, ,, Sun (♈S)	37° 9′	37° 9′
∴	,, Elongation (First length, SB)	11°27′	11°27′
	Subtract Parallax (see Table, p. 151.)	1°	58′
∴	**Second length** (SB_1)	**10°27′**	10°29′
B.	True Latitude of Moon (LB)	− 3°53′	− 3°53′
	Subtract Parallax (see Table, p. 151)	10′	11′
∴	**Second breadth** (L_1B_1)	**− 4°3′**	**−4°4′**

As true lunar longitude was 48°36′, we must, in accordance with Table on p. 168, subtract from the second length, $\frac{1}{4}$(− 3°53′) or $\frac{3}{10}$ (− 3°54′) i.e., we must add 1°1′, or 1°13′. We thus get

		Maimonides	Calculation
	Second length (SB_1)	10°27′	10°29′
	Add B_1C (המעגל)	1° 1′	1°13′
∴	**Third length** (SC)	**11°28′**	**11°42′**
∴	Ascension Difference RC = $\frac{6}{5}$ Third length	13°46′	14° 4′
	Subtract $\frac{2}{3}$ of LB (Maimonides) or $\frac{2}{3}$ L_1B_1 (Calculation)	2°35′	2°43′
∴	**Arc of Vision**	**11°11°**	**11°19′**

From equation Sin RH or Sin SP = ·848 Sin RA, it follows that when RA = 11°11′, Sin RH = ·848 × ·1939 = ·1644, whence **RH = 9°28′**, and when RA = 11°19′, Sin RH = ·848 × ·1965 = ·1663, whence **RH = 9°34′**. We thus see that by using Maimonides' data we obtain, by his method, a value of 9°28′ for RH, which is nearer the correct value of 8°42′ obtained by the direct method described on pp. 163 and 164 than is the value of 9°34′ obtained by using the more correct data. In other words, although the method described by Maimonides is not quite exact, he manages to obtain a fairly accurate result by using data which to a certain extent cancel out the inaccuracy of the method. (See p. 174)

NOTE. Because Maimonides obtained the values of:

Elongation > 11°, and Arc of Vision > 11°,

he comes to the conclusion that in accordance with his table (see p. 170) the moon would be visible.[1] As, however, the direct method of calculation shows that the moon was more than 1° below the horizon (LH = − 1°12′), it could not have been visible— if its longitude and latitude were those assigned to it by Maimonides.

Western Amplitude. The last thing with which the Calendar Council concerned themselves in the examination of the witnesses was the Western Amplitude of the moon, i.e., its horizontal angular distance from the west point, at the time in question[2] לאין היה נוטה. In other words, the witnesses were asked whether they saw the moon North or South of the Western point—and how far, approximately, it was from that point when it was setting at the Western horizon.

" To calculate this," says Maimonides rightly,[3] " it is first necessary to ascertain the moon's declination." For L_1D_1 (Fig. 52) being known, the spherical triangle L_1D_1A can be solved and the required arc AL_1 computed (A being the Western point, or point of intersection of the equator with the Western horizon). Thus:-

(1) $\text{Sin } L_1D_1 = \text{Sin } L_1B_1 \text{ Cos } 23\tfrac{1}{2}° + \text{Cos } L_1B_1 \text{ Sin } ♈B_1 \text{ Sin } 23\tfrac{1}{2}°$

(2) Having thus found Sin L_1D_1 we obtain, $\text{Sin } AL_1 = \dfrac{\text{Sin } L_1D_1}{\text{Cos } \phi}$

From this it follows that (i) when the moon is on the ecliptic, i.e., when $L_1D_1 = 0$, $AL_1 = 0$, i.e., the moon sets at the Western point, and (ii) the direction and magnitude of AL_1 depends upon those of L_1D_1; if L_1D_1 is positive (or North) then AL_1 is North, and vice-versa.

Thus for sunset of 2nd Iyar, 1178, we have.

$$L_1B_1 = -4°3' \text{ and } ♈B_1 = 47°36'$$

$$\therefore \quad (1) \quad \text{Sin } L_1D_1 = -\text{Sin } 4°3' \text{ Cos } 23\tfrac{1}{2}° + \text{Cos } 4°3' \text{ Sin } 47°36' \text{ Sin } 23\tfrac{1}{2}°$$

(since Cos (− 4°3′) = Cos 4°3′)

[1] Kidd. Hach. xvii. 22. Some texts say: Because the Elongation exceeded 10° It should be 11°.

[2] Rosh Hashanah, ii. 6. [3] Kidd. Hach. xix. 9.

$$= - \cdot 0707 \times \cdot 9171 + \cdot 9975 \times \cdot 7385 \times \cdot 3987$$
$$= - \cdot 0648 + \cdot 2937 = \cdot 2289$$
$$\therefore \quad L_1D_1 = 13°14'$$

(2) $$\text{Sin } AL_1 = \frac{\text{Sin } L_1D_1}{\text{Cos } 32'} = \frac{\cdot 2289}{\cdot 848} = \cdot 2652$$
$$\therefore \quad AL_1 = 15°23'$$

Maimonides considers it sufficient to estimate the magnitude of AL_1 very roughly, because the witnesses could not reasonably be expected to estimate, from mere inspection, with any degree of accuracy, the amount of this Western Amplitude AL_1. Hence Maimonides ignores parallax in this calculation and takes the moon's position as roughly: *longitude 49°, latitude-4°*. He further simplifies the calculation in the evaluation of the declination by ignoring lunar latitude altogether—since it is so small. The equation (1) therefore becomes:-

Sin BD = ·3987 Sin ♈B = ·3987 Sin 49° = ·3987 × ·7547 = ·3009 whence BD = 17°31′, say 18°.

This is much higher than the accurate value of 13°14′ obtained by means of the exact formula. Indeed, if we proceed to use this value (18°) for estimating the Western Amplitude under (2) we obtain

$$\text{Sin } AB = \frac{\cdot 3010}{\cdot 8480} = \cdot 3549, \text{ whence } AB = 20°47',$$

which is about 35% higher than the accurate value we obtained, viz., 15°23′. Maimonides, however, does not proceed in this way, but gives the following rough rule for estimating Western Amplitude:-

" Having found the moon's declination, we estimate the horizontal distance of the moon from the West point by taking the algebraical sum of the declination and latitude." Thus, declination being 18° (North), and latitude being − 4°3′ (South) the moon set at a point 18° − 4° = 14° North of the West point[1]—which is nearly the same as the accurate value 15°23′.

The following table compares the declination of the moon as found by means of the formula Sin BD = ·3987 Sin ♈B, with those given by Maimonides:-[2]

[1] Kidd. Hach. xix 9 & 10. [2] Ibid xix. 6.

Lunar Longitude	Lunar Declination	
	Calculated	Maimonides
10°	3°58′	4°
20°	7°50′	8°
30°	11°30′	11°30′
40°	14°51′	15°
50°	17°47′	18°
60°	20°12′	20°
70°	22° 0′	22°
80°	23° 7′	23°
09°	23°30′	23°30′

Maimonides forestalls any criticism of such inaccurate methods as the one just described by the following paragraphs in Chapter XI. of *Hilchoth Kiddush Hachodesh*: " Lest any mathematician think that the approximate methods sometimes described are due to my ignorance of the more accurate mathematical ones, let him dismiss such an idea from his mind. I have only employed such methods when I was convinced, by comparison with the results obtained by the strictly mathematical—but more laborious—methods, that the calculation of the moon's visibility is not appreciably affected by such approximation. He will further find that any appreciable positive or negative differences incidental to such approximate methods cancel themselves out, with the consequence that sufficiently accurate results are finally obtained without the labour of a long calculation which might frighten away the non-mathematical reader."

שמא יתבונן חכם...... ויראה קירוב מעט במקצת הדרכים ויעלה על דעתו שנתעלם ממנו דבר זה... אל יעלה זה על דעתו אלא כל דבר שלא דקדקנו בו מפני שידענו בעקרי הגימטריאות... שאין דבר זה מפסיד בידיעת הראיה ואין חוששין בו... וכן כשיראה בדרך מן הדרכים חסרון מעט מחשבון הראוי לאותה הדרך בכוונה עשינו זה לפי שיש כנגדו יתרון בדרך אחרת עד שיצא הדבר לאמתו בדרכים קרובים בלא חשבון ארוך כדי שלא יבהל האדם שאינו רגיל בדברים אלו ברוב החשבונות שאין מועילין בראית הירח.

Summary. We may now summarise Maimonides' somewhat complicated method of ascertaining whether the moon was visible on the evening of a given day, and how far it was from the West point.

(1) The mean position of the sun on that evening is calculated from the mean position at the epoch.

(2) The true position of the sun is then ascertained by adding (or subtracting) the appropriate equation of the centre.

(3) The mean position of the moon is then ascertained from its position at the epoch, and as a preliminary test the mean position of the sun is subtracted from it. The mean elongation thus obtained is an indication whether any further calculation is necessary or not—the minor and major limits being $2\frac{1}{2}°$ and 31°.

(4) The mean anomaly of the moon is also found from its mean anomaly at the epoch; and from a knowledge of the double elongation the amount of prosneusis is ascertained, which, when added to the mean anomaly, gives the true anomaly, from which the equation of the centre is calculated. The latter added to, or subtracted from, the mean longitude of the moon gives its true longitude.

(5) The difference between the lunar and solar true longitudes gives the true elongation—which forms another sign-post, inasmuch as no further calculation is needed unless this elongation lies between 9° and 24°.

(6) The mean longitude of the ascending node—calculated from its mean longitude at the epoch—subtracted from that of the moon gives the argument of the moon's latitude from which,

(7) The moon's latitude is calculated.

(8) The moon's true longitude and latitude enable one to find:

(i) The arc of vision (RA)

(ii The luni-solar vertical distance (RH)

(iii) The lunar declination

(iv) The lunar Western Amplitude.

Calculation of the of the Width of the Lunar Crescent. (Fig. 53). Let O be the centre of the moon; OE and OS, the directions of the earth and sun respectively. BAC and BDC are great circles perpendicular to OE and OS respectively.

Therefore, BDCD′ is the plane separating light from darkness.

and BACA′ is the plane separating the portion turned towards the observer from that turned away from him.

∴ Lune BACD constitutes the visible luminous portion.

As the inner semicircle BDC is seen obliquely, it is projected perpendicularly on ABC, into an ellipse BMC, whose semi-major axis OB is the radius of the moon, and whose semi-minor axis OM keeps on varying.

Now OM = OD Cos DOM = OA Cos SOV (since DOS = MOV = 90°)

∴ **AM** = OA − OA Cos SOV = **OA (1 − Cos SOV)**

and Arc of Crescent = $\frac{1}{2}\pi OA^2(1 - \text{Cos SOV})$

As the distances of the moon and earth from the sun are very large, OS and ES may be taken as parallel, and therefore angle SOV = OES = lunar elongation.

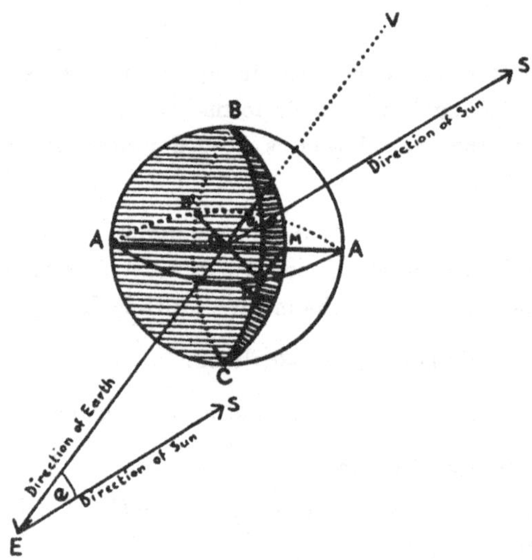

Fig. 53

∴ **Width (AM) of Crescent = OA(1 − Cos Elongation)**

Thus, in our case, Elongation OES = 11°27′, whose Cosine = 0·9801.

∴ Width of Crescent = OA(1 − ·9801) = about $\frac{1}{50}$ of the radius of the moon.

The smallest visible crescent is that corresponding to a true elongation of 9° (p. 160), and its width = (1 − Cos 9°)OA = (1− ·9877)OA

= about $\frac{1}{80}$ the radius of the moon. Israeli[1] gives the width of the crescent for 9° elongation as $\frac{1}{18}$ of the diameter, or $\frac{1}{9}$ of the radius, of the moon (ב' שלישי אצבע במדה שבה הוא שטח גופו י"ב אצבעות) a width that rightly corresponds to an elongation of 27° (1 − Cos 27° = 1 − ·8910 = $\frac{1}{9}$).

NOTE: The ratio of the width of the visible illuminated portion to the whole disc is called the *phase* of the moon.

The following Table gives the widths of the crescent for different elongations, in relation to the semi-diameter OA.

Elongation (E)	Width of Crescent (1 − Cos E) OA
9° (see p. 160)	(1 − ·9877) OA = $\frac{1}{80}$ OA
15° (see p. 160)	(1 − ·9659) OA = $\frac{1}{30}$ OA
24° (see p. 161)	(1 − ·9135) OA = $\frac{1}{12}$ OA
45° (octants)	(1 − ·7071) OA = $\frac{3}{10}$ OA
90° (quadrants)	(1 − 0) OA = OA (half moon)
180° (opposition)	(1 + 1) OA = 2OA (full moon)

[1] Yesod Olam, iii, 15.

CHAPTER XVI.

The Examination of Witnesses.

על פי שני עדים יקום דבר (דברים י"ט, ט"ו).

"At the mouth of two Witnesses shall the matter be established".

(Deut. xix. 15)

The Calendar Council (סוד העבור)[1] at first consisted of priests[2] who examined the evidence of witnesses who reported having seen the new moon. At a later period, the priests were replaced by three members of the Synhedrin,[3] who were specially qualified by their mathematical and astronomical skill, and one of whom was the Nasi—or President—of the Synhedrin. In the absence of unanimity, the Court was increased to five, and, if necessary, to seven members.[3] In addition to their duty to examine witnesses, the Calendar Council also had to decide which year was to be made a leap year (or embolistic year). The question was originally decided by the state of the crop.[4] For, since the lunar year—consisting as it does of 12 lunar months, i.e., of 12 lunations of average duration of about 29½ days—lasts 354 days (see p. 186), whilst the solar year lasts about 365 days, the lunar is shorter than the solar year by about 11 days. After about 3 years, therefore, there is a discrepancy of more than a month between the two kinds of year, so that Passover, for instance, occurring as it does in Nisan—which should be the first Spring month חדש האביב (Deut. xvi. 1)—would, after a few years, actually occur in the Winter. Hence, if in any year the crop was found to be still too unripe to allow the Omar to be offered at the proper time (16th Nisan), the Calendar Council intercalated an extra month between Adar and Nisan—although it is also recorded[5] that an extra month was sometimes inserted between Ab and Elul instead of between Adar and Nisan.

At a later period the actual position of the sun in the ecliptic was also taken into account, and if observation, or calculation,

[1] Kethuboth 112a. [2] Rosh Hashanah, i. 7. [3] Synhedrin, i. 2.
[4] Cf. Synhedrin 11a & b. [5] Synhedrin 12a.

showed that the vernal equinox would fall after the 16th of Nisan, a leap year was declared[1] כד חזית דמשכה תקופת טבת עד שהסר בניסן עברה להאי שתא.

The extra month consisted of 29 days[2] אדר הסמוך לניסן לעולם. חסר. It was owing to this particular function of intercalation that the King and High Priest were excluded from membership of the Calendar Council: the King might be biassed in favour of too many leap years in order to save expenditure in connection with the Army, which is paid by the year; whilst the High Priest, who has to immerse his body in spring water (*Mikvah*) 5 times on the Day of Atonement, might oppose intercalation in order to bring about the falling of that day during the warm period of the year[3] אין מושיבין מלך וכ"ג בעיבור שנה... מלך משום אסניא, כ"ג משום צינה.

The Council, as we saw, was called *Sod Haibbur* (סוד העיבור) but this name was also given to the secret method in use for calculating whether an extra month was, or was not, necessary to be intercalated[4] (i.e., the method of calculating the time of occurrence of an equinox—see p. 85).

The Council sat in Jerusalem at a court called *Beth Yaazek* (בית יעזק) on the 30th of each month, to receive witnesses. Rabbi also established another court in a southern province *Ain Tab* (עין טב)[5]. If after cross-examination (see p. 180) and further confirmation of their evidence by mathematical calculation (in the manner described in the foregoing pages), the Council concluded that the new moon was, indeed, seen by the witnesses at the time they mentioned, that day was declared to be the first of the new month (ראש החדש). If no witnesses arrived by the 30th day, then the 31st day was declared to be the first of the month without any further waiting. If, therefore, people who saw the new moon on the 29th or 30th day, lived more than two days', or one day's journey, respectively, from the Calendar Council, there was no need for them to come to attest (see Kidd. Hach. iii. 1).

A month of 29 days was called a defective month (חדש חסר), and one of 30 days was a full month (חדש מלא or חדש מעובר). It was

Rosh Hashanah, 21a. [2] Rosh Hasharah 19b.
[3] Synhedrin 18b. (see Rasbi ad loc). [4] Rosh Hashanah 20b.
[5] Rosh Hashanah 25a.

possible for 3 months to be consecutively full if—owing to unfavourable atmospheric conditions—the new moon was not visible. If, however, no witnesses appeared for 3 months, then the beginning of the next month was determined by calculation alone. No year was allowed to have more than 8, or less than 4, full months,[1] so that no lunar year lasted more than 356 or less than 352 days.

The respect in which the Calendar Council was held is illustrated by the following quotation, showing how R. Eleazar considered himself privileged to have been chosen a member thereof:[2] "R. Eleazar, when he moved to Palestine said, 'I have been saved from one of the curses with which God cursed the false prophets, viz., that they shall not enter the land of Israel (Ezekiel xiii, 9). When he received his rabbinate, he said that he was saved from the 2nd curse, i.e. not to be inscribed in the writing of the house of Israel (Ibid.). When he was elected to a seat on the Calendar Council, he considered himself saved from the 3rd curse, not to be in the assembly of God's chosen people (Ibid.)"

ר׳ אלעזר כי הוה סליק לא"י אמר פלטי לי מחדא, כי סמכוהו אמר פלטי לי מתרתי כי אותבוהו בסוד העבור אמר פלטי לי מתלת, וכו׳.

It would appear that the Council had mathematical specialists to act as their assessors (אילין דמחשבין)[3]

Character of the Witnesses. No evidence was considered valid unless it was given by at least two reputable male witnesses; gamblers, usurers, slaves, and others, whose reputation was not beyond reproach, were disqualified. The reason was that certain evil-minded people, in order to upset the Jewish Calendar, sometimes bribed people deliberately to deceive the Court.[4] At a later date R. Judah Hanasi allowed criminals to give evidence before the Calendar Council, although they would have been disqualified as witnesses in a Court of Justice. He also accepted 'hearsay' evidence[5]—thus showing that he practically entirely relied upon calculation alone רבי ביטל את המשואות והתיר רוצח והתיר עד מפי עד.

Examination of the Witnesses. It was a religious duty for any one who saw the new moon, and who lived within 24 hours' journey of Jerusalem, to come and testify; as an encouragement, such witnesses were entertained to a sumptuous meal and given certain privileges. Questions were asked of them regarding the position of the moon in relation to the sun—whether East or West of it; whether North or South of the West point; its

[1] Erachin, ii. 3. [2] Kethuboth 112a.
[3] Jer. Succah, iv. 1. [4] Rosh Hashanah ii. 1. [5] Jer. Rosh Hashanah, ii. 1.

altitude and azimuth; the direction of the crescent's horns; the width of the crescent, etc.[1] כיצד ראית את הלבנה, לפני החמה או לאחר החמה לצפונה או לדרומה, כמה היה גבוה ולאין היה נוטה וכמה היה רוחב.

Position of Moon in relation to the Sun. Since the moon and sun travel in the same direction from West to East and the former gets ahead of the latter at the rate of about 12⅛° a day, it follows that, starting from a conjunction, the new moon will be ahead of the sun לפני החמה (i.e., east of it) until the elongation is 180° (at full moon), after which it will be west of the sun לאחר החמה (i.e., behind it). But the diurnal rotation of the heavens being from East to West, it is clear that the *new moon*—being East of the Sun *will appear at the eastern horizon after sunrise, and will set at the western horizon after sunset*; the reverse being the case with the waning moon. Hence—as the above passage concludes—" if the witness stated that he saw the moon sink at the western horizon before sunset, his evidence is worthless " אמר לפני החמה לא אמר כלום.

For the same reason evidence to the effect that the moon was seen to rise in the morning before sunrise and to set in the evening after sunset could not be reliable (see pp. 9 and 10), because it must necessarily refer to the waning moon in the morning and the new moon in the evening. For, as we saw on p. 160, for the moon to be visible at all, its true elongation must be not less than 9°; but at the rate at which elongation increases—viz., about ½° per hour—it must take about 18 hours on either side of a true conjunction for the moon to cover such an arc, so that an interval of about 36 hours must elapse between the disappearance of the crescent of the old and the appearance of that of the new moon. R. Zera, in the name of R. Nachman gave the minimum period of invisibility of the moon as 24 hours[2]. כ"ד שעי מיכסי סיהרא.

Was the moon North or South of the Sun? When the moon's latitude is positive, the moon is of course North of the sun (so that the witness facing the western horizon should have seen the moon to the right of the sun). When the latitude is zero (i.e., the moon is at a node in the ecliptic), the moon and sun are seen to set at the same point of the horizon. When the

Rosh Hashanah, ii. 6.

Ibid, 20b. Maimonides (Kidd. Hach. i. 3.) gives the period of invisibility of the moon as *about* 2 days.

latitude is negative, the moon is seen south (i.e., to the left) of the sun.

According to one anonymous rabbi, the moon is always North of the sun, so that if a witness says that he saw it South of the sun his evidence is valueless (לצפונה דבריו קיימין. לדרומה לא אמר כלום), whilst according to another anonymous authority the opposite is the case. The Talmud reconciles the two statements by saying that in the summer the moon is North of the sun, whilst in winter it is South of the sun[1] (כאן בימות חחמה כאן בימות הגשמים). Rashi gives a long explanation of why this should be so. As a matter of fact, all these statements, including Rashi's explanation, are incorrect. It is true—as Rashi points out, and as we saw on p. 70—that between the spring and autumn equinoxes the sun is North of the equator (and therefore rises North of East and sets North of West), whilst between the autumn and spring equinoxes, it is South of the equator (and therefore rises South of East and sets South of West), but this gives no information as to whether during these periods the moon will be North or West of the sun; for the latitude of the moon depending, as we have seen it to do, on the movement and position of the node (see p. 147) has no relation to the seasons.

The Altitude and Amplitude of the Moon. The manner of computing these has been described on pp. 162 and 172. The Talmud says,[2] that if one witness gives the altitude as two *Mardeoth* (מרדעות) and the other gives it as three, their evidence is not necessarily inconsistent. If, however, there is a difference of two *Mardeoth* (one saying 3 and the other 5), the discrepancy is too great to be overlooked אחר אומר גבוה ב׳ מרדעות, ואחר אומר ג׳ עדותן קיימת, וכו׳.

It is not quite clear what altitude is indicated by a Mardeah (מרדעת). The word as it stands means an "Ox-Goad", but it is clear it cannot have that meaning in this passage. Maimonides[3] translates it as "unit of altitude" (קומה). It is possible that the word may be a corruption of (מדרגה), and is therefore aptly to be translated by "degree".

Direction of the Horns of the Crescent. As the moon receives its light from the sun, it is obvious that the horns of the

[1] Rosh Hashanah, 24a. [2] Ibid. [3] Kidd. Hach., ii. 5.

crescent must always point away from the sun, so that in the language of R. Yochanan " the sun has never seen the concavity of the lunar crescent "[1] מעולם לא ראתה חמה פגימתה של לבנה. Hence, when the new moon is seen above the horizon after sunset its horns must point away from the horizon. On the other hand, the horns of the waning moon will point towards the horizon—i.e., also away from the sun (see Fig. 54).

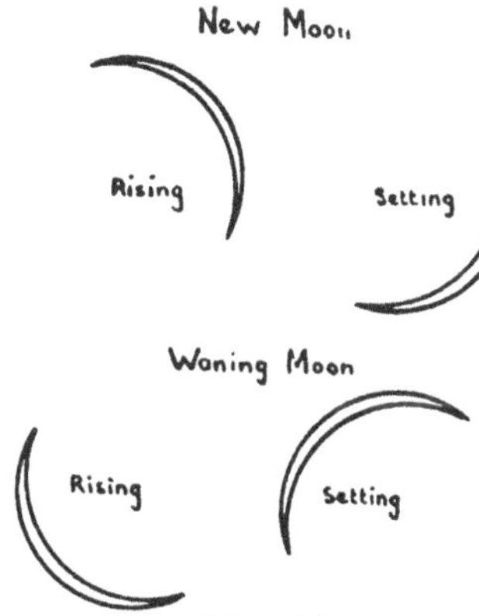

Fig. 54

The Width of the Crescent. The manner of calculating this has been described on p. 175. R. Gamliel, as we saw on p. 9, had charts of the moon's crescent in different phases, and he used to ask witnesses to point to the particular phase resembling the crescent they stated to have seen[2] דמות צורות לבנה היו לו לר"ג בטבלא ובכותל בעלייתו שבהן מראה את ההדיוטות ואומר הכזה ראית או כזה.

Dissemination of the News. When the Court was satisfied that the new moon had actually been seen, the President declared the new moon to be consecrated[3] and the news was spread to people outside Jerusalem by means of bonfires, or beacons, on the tops of mountains. When the awaited signal was observed at neighbouring mountains, similar fire signals were lit, and thus the information was transmitted to distant places. This method was continued until a deliberate attempt to confuse the Jews was made by the Cuthites, (Samaritans), who maliciously lit signals at improper times.[4] Since then these fire signals were altogether

1 Rosh Hashanah, 23b. 2 Ibid, 24a. 3 Ibid, ii. 7.
4 Ibid, ii. 2. It is probable that the word כותי (Samaritan) in the text is the Censor's substitute for the word גוי (Gentile), which was believed to discredit Christianity.

abolished by R. Judah Hanasi (towards the end of the 2nd century) and messengers or couriers, to convey the tidings, were sent out instead.[1] As these could not reach the more distant places before the 31st day of the old month, the residents of those places used to celebrate two days for their holidays[2] (יו"ט שני של גליות)—because they were not sure whether the first of the month was on the 30th or on the 31st day.

In the case of New Year, two days were kept even in Jerusalem (and are still so kept) in case witnesses arrived late in the day.[3]

These couriers were at first sent out only 6 times a year, viz., Nisan—on account of Passover; Ab—on account of the Fast on the 9th; Iyar—on account of a minor feast, Purim Katan; Elul, on account of the New Year starting on the first of the following month; Tishri, in order to fix the order of the festivals; Kislev, on account of the feast of lights—Chanucah. Rabbi sent out couriers also in Adar—on account of Purim.[4] על ששה חדשים השלוחין יוצאין... רבי התקין שיוצאין אף על אדר מפני פורים.

Adolf Schwarz quotes this Tosefta as proof that the substitution of couriers for fire signals is not due to Rabbi—to whom the Palestinian Talmud, quoted above, attributes this change. Further, Schwarz makes the plausible suggestion that Rabbi could not pay much attention to the evidence of witnesses, but must have relied mainly on his mathematical calculations, because, on all the months other than Nisan and Tishri, the couriers left Jerusalem before the examination of the witnesses was completed. In any case, under the despotic rule of the Romans, witnesses had to be dispensed with, and reliance was placed upon mathematical methods alone.

[1] Jer. Rosh Hashanah, ii. 2. רבי בטל את המשואות [2] Pesachim, 52a.

[3] Kidd. Hach. v. 7 & 8. [4] Tosefta Rosh Hashanah, 1. quoted by A. Schwarz, Der jüd. Kalender. Breslau, 1872, p. 28.

CHAPTER XVII.

The Fixed Calendar.

הלכה למ"מ היא שבזמן דאיכא סנהדרין קובעין על פי הראיה ובזמן שאין סנהדרין קובעין ע"פ החשבון שאנו מחשבין בו היום. (הלכות קדוש החדש ה', ב').

"It was divinely ordained to Moses on the Mount Sinai that when there is a Jewish Synhedrion, the month is determined by the Phase method, but in the absence of a Synhedrion the Fixed Calendar now in use must be employed."

(Maimonides, Kiddush Hachodesh v. 2.)

The Phase Method of determining the beginning of a month described in the preceding pages, prevailed until the time of Abbaye and Raba[1] (middle of 4th century), when it was replaced by the **Fixed Calendar Method** which makes use of a *Mean Conjunction* or *Molad* to determine the beginning of a month. The interval between two consecutive mean conjunctions, (a lunation or mean synodic month), being of constant length, viz., 29d. 12h. 44m. $3\frac{1}{3}$s., it at once disposes of the need of laborious calculations every month, and of the need of disseminating the news each time —because the beginning of any month which is n months distant from some epoch whose Molad is known, can be easily computed by anybody from the equation :

$$\textit{Required Molad} = \textit{Molad at epoch} + n\ (29d.\ 12h.\ 44m.\ 3\tfrac{1}{3}s.)$$

It might at first sight appear that as the Phase Method enabled one to ascertain the exact moment of a true conjunction, it was much more exact than the Fixed Calendar Method which relies upon averages. This, however, is not so, because the beginning of the month was not fixed by the moment of the true conjunction but by the moment the crescent of the new moon was seen—which was *at least 18 hours* after a conjunction (see p. 177). On the other hand, the interval between a true and a mean conjunction is *at most 15 hours*—which may be positive or negative (see p. 146). In one respect, however, the Phase Method was more correct because the lunar was made to keep pace with the solar year by intercalating an extra month whenever there was actual need for it, (as found by the size of the discrepancy between the date of Passover and that of the calculated, or observed, vernal equinox) instead of intercalating at regular intervals — as is done under the Fixed Calendar Method. On the whole, however, the advantages of the Fixed Calendar Method considerably outweigh its disadvantages.

[1] Kidd. Hach. v. 3.

The Principles of the Fixed Calendar. As the average length of a synodic month is about 29½ days, and as a calendar month must have an integral number of days and must start at the same hour of the day — viz., 6 p.m. — it has been agreed to make a civil month consist alternately of 30 days (=*full month* חדש מלא or חדש מעובר) and 29 days (=*defective month* חדש חסר)[1] as follows:-

Months of 30 days	(1) Nisan	(3) Sivan	(5) Ab	(7) Tishri	(9) Kislev	(11) Shebat	Total = 180
Months of 29 days	(2) Iyar	(4) Tammuz	(6) Elul	(8) Mar Cheshvan	(10) Tebeth	(12) Adar	Total = 174

Such an arrangement of alternate full and defective months constitutes what is called an *ordinary regular year* (שנה פשוטה כסדרה), consisting of 180 + 174 = 354 days = 50 weeks + 4 days.[2]

Modifications. This arrangement must, however, be modified in a number of ways, as follows:

1. Disparity between the ordinary regular and astronomical lunar year.

If a lunation were to last exactly 29½ days, then an astronomical lunar year consisting of 12 lunations would contain exactly 354 days, and the two kinds of years would agree perfectly. As, however, the length of a lunation is 29d. 12h. 44m. 3⅓s., the astronomical lunar year exceeds the civil lunar year by 12 × (44m. 3⅓s.), i.e., by 8h. 48m. 40s., or over ⅓ day. Hence it is necessary, at fixed intervals, to add a day to the ordinary year by making one of the defective months full, giving rise to an *ordinary full year* שנה) (פשוטה שלמה of 355 days. The defective month chosen for the purpose is Marcheshvan, so that *in a full year, Tishri, Marcheshvan and Kislev instead of having 29 and 30 days respectively all have 30 days.* Such a procedure, however, if adopted say every 3rd year, necessarily overbalances the disparity, and hence at other fixed intervals, the excess so introduced is adjusted by subtracting a day from one of the full months, making that particular year contain only 353 days. Such a year is called an *ordinary defective year* שנה פשוטה חסרה,

[1] Rosh Hashanah 19b.

[2] Erachin 9b. אין בין ראש השנה לראש השנה אלא ד' ימים, ואם היתה שנה מעוברת חמשה.

and the full month chosen for the purpose is Kislev, so that *in a defective year, Marcheshvan, Kislev and Tebeth all have 29 days instead of* 29, 30 *and* 29 *days respectively.* The arrangement is shown in the following Table :-

Name of Month	Ordinary year: Regular	Ordinary year: Defective	Ordinary year: Full	Leap year: Regular	Leap year: Defective	Leap year: Full
Tishri	30	30	30	30	30	30
Marcheshvan	29	29	30	29	29	30
Kislev	30	29	30	30	29	30
Tebeth	29	29	29	29	29	29
Sh'bat	30	30	30	30	30	30
Adar	29	29	29	30	30	30
V'Adar	—	—	—	29	29	29
	177d = 25w + 2d	176d = 25w + 1d	178d = 2[illegible] w + 3d	207d = 29w + 4d	206d = 29w + 3d	208d = 29w + 5d
Nisan	30	30	30	30	30	30
Iyar	29	29	29	29	29	29
Sivan	30	30	30	30	30	30
Tammuz	29	29	29	29	29	29
Ab	30	30	30	30	30	30
Elul	29	29	29	29	29	29
	177d = 25w + 2d	177d = 25w + 2d	177d = 25w + 2d	177d = 25w + 2d	177d = 25w + 2d	177d = 25w + 2d
Total days or weeks in the year.	354d = 50w + 4d	353d = 50w + 3d	355d = 50w + 5d	384d = 54w + 6d	383d = 54w + 5d	835d = 55w

NOTE. Marcheshvan and Kislev are the two months chosen for the purpose because amongst the Babylonians (from whom the Jews borrowed the names o the months), these two months contained sometimes 29 and sometimes 30 days.

2. Disparity between the Astronomical Lunar and Astronomical Solar year.

A lunar year of 12 lunations consists of 354d. 8h. 48m. 40s., whilst R. Adda's solar year consists of 365d. 5h. 55m. 25$\frac{25}{57}$ s. Therefore the lunar, falls short of the solar, year by 10d. 21h. 6m. 45$\frac{25}{57}$s. i.e, by $\frac{\text{10d. 21h. 6m. } 45\frac{25}{57}\text{s.}}{\text{29d. 12h. 44m. } 3\frac{1}{3}\text{s.}} = \frac{53580310}{19 \times 7654330}$

$=$exactly $\frac{7}{19}$ of a synodic month.

Hence, if in the course of every cycle of 19 years we intercalate 7 extra months, making 12 ordinary years of 12 months each (=144 lunations), and 7 leap years of 13 months each (=91 lunations), we shall get the following results :-

19 Solar (R. Adda's) years = 19(365d. 5h. 55m. $25\frac{25}{57}$s.)
= **6939d. 16h. 33m. $3\frac{1}{3}$s.***

Consisting of **235 lunations** = 235(29d. 12h. 44m. $3\frac{1}{3}$s.)
= **6939d. 16h. 33m. $3\frac{1}{3}$s.**

so that, at the end of the cycle, the sun and moon will have exactly the same relative mean positions in the heavens as at its beginning (on the assumption that R. Adda's year represents the exact duration of a tropical year).

The Nineteen Years' Cycle called the **Small Cycle** (מחזור קטן) to distinguish it from the **Twenty-Eight Year Cycle** or Big Solar Cycle ((מחזור חמה or מחזור גדול) to be mentioned later, (p. 199), was first introduced by the Greek astronomer Meton in the 5th century B.C., who, taking the length of the year as $365\frac{1}{4}$ days (the same as Samuel's year), discovered that 19 solar years expressed in days, hours, etc., contain *almost* exactly 235 lunations. R. Adda's year is, therefore, apparently an *arbitrary* figure designed to make the solar year equal to *exactly* $\frac{1}{19}$ of 235 lunations. R. Adda's year is nowhere specifically mentioned in the Talmud, and it is, therefore, questionable whether that figure is really due to R. Adda bar Ahaba, the contemporary of Samuel. It is clear, however, that whoever introduced it was acquainted with the Metonic Cycle.

Reference has already been made on p. 75, to the arguments in favour of and against the contention that R. Adda's t'kufah was known to the Talmudic Rabbis. Additional points in the controversy will be found in the following works: Ch. Z. Slonimsky, *Yesode Ha-Ibbur*, 1865; A Schwarz, *Der jüd. Kalender*, 1872; E. Mahler, *Handb. d. jüd. Chronologie*, 1916. See also pp. 202–204 below.

The origin of the division of the Chelek into 76 Regaim giving 82,080 regaim to the hour is now clear. For, since 76 = 19 × 4, all fractions are avoided in the arithmetical computations of cycles. Thus, R. Adda's year is given in the form of 365d. 5h. 997ch. 48reg.; and therefore a nineteen years' cycle contains 693d. 16h. 595ch., which is equal to 235 synodic months of 29d. 12h. 793ch. This division of the chelek is apparently a post-Talmudic one, for the Talmud gives the number of regaim to the hour as 58,888.[1]

* This is the same as 6939d. 16h. 595ch

[1] Berachoth, 7a.

The Order of the Leap Years is as follows: 3, 6, 8, 11, 14, 17, and 19; the numbers indicating the place of the year in the cycle. The Hebrew mnemonic is גו"ח אדז"ט (*Guchadz̤t*). Thus the year 5689 W.E. is a leap year, because $\frac{5689}{19} = 299 + \frac{8}{19}$ which means that the year 5689 is the **8th** year in the 300th cycle.

Calculation of the Molad for any given month. תיקון המולדות (see Kidd. Hach. vi)

The Jewish month begins at 6 p.m. of the day on which the moon is in conjunction. Hence, if the Molad of any given month is known, that of the next month is ascertainable by adding an average lunation, viz., 29d. 12h. 44m. $3\frac{1}{3}$s., or—which is the same thing—1d. 12h. 44m. $3\frac{1}{3}$s., since *29 days* after a certain event will fall on the same day of the week as *1 day* after the event.

The figure **1d. 12h. 44m. $3\frac{1}{3}$s.** or **1d. 12h. 793 Chelakim** (a *Chelek* being $\frac{1}{1080}$ hr.—see p. 101) is therefore called the **Character of a month.** (שארית חדש הלבנה)

Hence, if the Molad of any given month is known, say $= M_1$ then that of any other month (M_2), *n* months after or before the given month is found from the formula:

$$M_2 = M_1 \pm n(1d.\ 12h.\ 44m.\ 3\tfrac{1}{3}s.) = M_1 \pm n(1d.\ 12h.\ 793ch.)$$

The **Epoch** or moment of the Molad from which all calculations are made, is the Molad of the Tishri preceding the month of Nisan in which, according to tradition, the world was created[1]

As the creation of the world is fixed by the Jews to have taken place in the year 3761 B.C.,* calculation backwards from the moment of an observed true conjunction, by means of the above formula places the Molad of that Nisan at Wednesday, $35\frac{2}{3}$m. past 3 a.m., i.e., 4d. 9h. 35m. 40s. (since 3 a.m. is 9 hours after 6 p.m.) Therefore the Molad of the preceding Tishri (i.e., 6 months earlier), is given by the formula:

$$M_{Tishri} = 4d.\ 9h.\ 35m.\ 40s. - \underbrace{6(1d.\ 12h.\ 44m.\ 3\tfrac{1}{3}s.)}_{9d.\ 4h.\ 24m.\ 20s.}$$

* From Biblical, Talmudic and other traditional data the period between Adam and the Exodus amounted to 2448 years and between the latter and the destruction of the 2nd Temple there elapsed another 1380 years, making a total of 3828 years; and as the destruction of the temple was considered to have taken place in the year 68 C.E., instead of 70 C.E., it follows that the 1st year of the W.E. Corresponded to 3761 B.C.(= -3760 Astronomical).

[1] Rosh Hashanah 11a. בניסן נברא העולם.

=4d. 9h. 35m. 40s. – 2d. 4h. 24m. 20s. (by eliminating 7 days)

=2d. 5h. 11m. 20s. or **2d. 5h. 204ch.** (ב"ה"ר"ד = **Baharad**)

This Molad is therefore called *Molad B$_a$h$_a$r$_a$d.* Also as the epoch is an imaginary one, having occurred *before* the assumed creation of the world, it is also called *Molad Tohu* (מולד תוהו) or *Imaginary Molad*

Simplification of Calculation.

It is obvious that if one had to calculate every time the number of months that have elapsed between the epoch and the month in question and then multiply that number by the " character " of a month, the work would become exceedingly laborious and tedious. Hence, the computation is simplified in the following manner :-

(i) A 19 year's cycle contains exactly 235 lunations equal to 6939d. 16h. 33m. 3⅓s (see p. 188). Eliminating the 991 complete weeks (=6937d.), we are left with the **Character of a Cycle** (שארית המחזור) viz. **2d. 16h. 33m. 3⅓s.** or **2d. 16h. 595ch.** As the number of cycles intervening between the Epoch and any given year is found by dividing that year (as expressed by the figures representing its place in the world era) by 19, the quotient thus found, multiplied by the character of the cycle, when added to the epoch, represents the Molad of Tishri of the first year of the cycle in which the given year occurs.

Thus the year 5689 when divided by 19 gives the quotient of 299 and remainder of 8, which means that 299 complete cycles have elapsed since the Molad Tishri, and that the year 5689 is the 8th year in the current 300th cycle. The first year of the current cycle was, therefore, 5682 and the Molad for Tishri in that year =2d. 5h. 11m. 20s. + 299(2d. 16h. 33m. 3⅓s.) which after multiplication, and the casting out of multiples of complete weeks, becomes **1d. 9h. 54m. 56⅔s.**

(ii) Having found the Molad for Tishri of the first year of the current *cycle*. the Molad of Tishri of the current *year* is computed as follows :

(a) Between the beginning of 5682 and the beginning of 5689, seven years have elapsed, two of which, the 3rd and 6th, (viz. 5684 and 5687) were leap years (see p. 189) and the other five were ordinary years. Now a leap year consisting of 13 lunations=

13(29d. 12h. 44m. 3⅓s.) which after eliminating multiples of complete weeks = **5d. 21h. 32m. 43⅓s.** – the **character of a leap year** (שארית שנה מעוברת) Similarly the **character of an ordinary year** of 12 lunations = **4d 8h. 48m. 40s.** (שארית שנה פשוטה)

(b) Therefore the Molad for Tishri of the current year 5689 = 1d. 9h. 54m. 56⅔s. + 5(4d. 8h. 48m. 40s.) + 2(5d. 21h. 32m. 43⅓s.) **= 7d. 1h. 3m. 43⅓s,*(= 7d. 1h. 67ch.)**

(iii) Having thus found the Molad for Tishri of the current year, the Molad for any given month is now ascertained by adding n(1d. 12h. 44m. 3⅓s.). where n is the number of lunations that have elapsed between the beginning of the year and the Molad in question.

Thus for the Molad of Tammuz 5689, for instance, we have, (since 5689 is a leap year), Molad Tammuz = 7d. 1h. 3m. 43⅓s. + 10(1d. 12h. 44m. 3⅓s.) = 1d. 8h. 24m. 16⅔s.

The general formula for computing the Molad (M_X) for any month in any given year is:

M_X = 2d. 5h. 11m. 20s. + Q(2d. 16h. 33m. 3⅓s.) + n_o (4d. 8h. 48m. 40s.) + n_l (5d. 21h. 32m. 43⅓s.) + n(1d. 12h. 44m. 3⅓s.),

where Q = number of complete cycles since the epoch,

n_o = number of ordinary years } since the beginning of
n_l = number of leap years } the current cycle

n = number of months that have elapsed since the beginning of the current year.

Postponements of the First Day of Tishri (Dechiyoth דחיות)

There are five separate occasions which necessitate the postponement of the New Year's Day by one or even two days. These postponements depending on the exact moment at which a Molad occurs, necessarily entail a lengthening of the year by one or two days (with the consequent lengthening of one or two months in that year) and are called **Dechiyoth** (דחיות) They are as follows: (See Kidd. Hach. vii)

1. **Dechiyath** c^h (דחית י"ח) If the Molad of Tishri occurs at, or past, noon (i.e., 18h. ח"י) of any day, New Year (ראש השנה) is postponed to the following day. Such a Molad is

* If after eliminating multiples of 7, there are no days left at all, 7d. is put in to denote that the Molad occurs on the 7th day in the week.

called an Old Molad (מולד זקן), because on the day of the new month, the moon is already old. It is not easy to understand the reason for this postponement. The explanation generally given is that with a mean conjunction at noon, the crescent of the new moon would not be large enough to be visible at the following sunset, and the Calendar Council, (when the Phase Method of Calendar making was in use), would, therefore, not have declared a new month on that day. As, however, Tishri is near the Autumnal Equinox when sunset is at 6 p.m. and as, at the rate of increase of elongation of about ½° per hour, (p. 122), the minimum mean elongation compatible with visibility, viz., 2½° (see p. 144) would be reached about one hour before sunset, it follows that even if the Molad occurred at 1 p.m. the moon would still be visible at sunset. It must therefore be assumed that those who instituted this dechiyah must have considered the minimum mean elongation limit to be 3° instead of 2½°.

Talmudic authority for this *dechiyah*, is claimed from the following rule given by R. Zera : " The time of a Conjunction is calculated : if it is found to be before 12 o'clock, then one can be certain that the crescent would be visible at about the time of sunset, but if it occurs after 12 o'clock, it is equally certain that the new moon would not be visible at about sunset."[1]

מחשבין את תולדתו, נולד קודם חצות בידוע שנראה סמוך לשקיעת החמה לא נולד קודם חצות בידוע שלא נראה סמוך לשקיעת החמה.

Notwithstanding, however, the apparently complete parallelism between R. Zera's rule and the *dechiyath yach*, it is practically certain that the similarity between the two is apparent rather than real, and is brought about by the word חצות, which means " half " or " middle " i.e. 12 o'clock. For, as the Calendar Council fixed a new month by the time of a *true* conjunction, the quoted passage cannot refer to a *mean* conjunction. On the other hand, if it refers to a true conjunction, then for visibility there must be a minimum *true* elongation of 9° (see p.170), and therefore חצות cannot signify 12 o'clock *noon*,—as Rashi translates it—since we must have an interval of 18 hours to produce an elongation of 9°. The conclusion, therefore, seems inevitable that the word חצות must be translated as 12 o'clock *midnight*, which would make the interval between the conjunction and sunset exactly 18 hours—just enough to increase the elongation to 9°—on the assumption that during the comparatively short interval of 18 hours, *true* elongation increases at the same rate as *mean* elongation.

As to why a celebrated astronomer like Samuel, who claimed to be able to fix a Calendar for the whole of the diaspora[2] and to whom the topography of the heavens was as familiar as that of his native town of Nehardea[3], should have ex-

[1] Rosh Hashanah 20a.
[2] Rosh Hashanah 20b יכילנא לתקוני לכולה גולה
[3] Berachoth 58b נהירין לי שבילי דשמיא כשבילי דנהרדעא

pressed ignorance of this rule, see (p. 12) can, in the author's opinion, be explained on the supposition that Samuel's answer to Aba may have been more laconic than sincere, wishing to imply that as an expert astronomer, he was not interested in a rule which is only correct on the average; for, as the rate of increase of true elongation is not constant, the rule can only be approximately correct. In any case, it is fairly clear that the *dechiyath Yach* cannot be correlated with R. Zera's statement—although by a curious coincidence depending on the meaning of the word חצות, the two rules seem to identical. Israeli[1] gives the minimum interval between a true conjunction and a visible crescent as 18 hours. He states that the width of the crescent would then be $\frac{1}{18}$ of the width of the moon. The latter statement is incorrect. (see p. 177.)

2. Dechiyath Adu (דחית אד"ו) If the Molad of Tishri occurs on Sunday, Wednesday or Friday, (the first, fourth or sixth days in the week), New Year's Day must be postponed (לא אד"ו ראש). Wednesday and Friday are excluded in order to obviate the occurrence of the Day of Atonement (10th Tishri) on Friday or Sunday, as in such a case great inconvenience and insanitary consequences, might follow from having two Sabbath days, (the Day of Atonement being the Sabbath of Sabbaths שבת שבתון), coming consecutively, as vegetables intended for food might decompose in two days, as would also dead bodies remaining unburied for the same length of time (משום ירקיא ומתיא)[2] Sunday is excluded to avoid the 7th day of Tabernacles (הושענא רבא) which occurs on the 21st of Tishri (and on which day the Willow twigs have to be beaten) falling on Saturday.[3] In these cases, therefore, New Year's Day is put off to the following day. As A = א = 1; D = ד = 4 and U = ו = 6, this postponement is called *Dechiyath Adu.*

From the rule לא אד"ו ראש, it follows that the first day of Passover (15th Nisan) cannot fall on Monday, Wednesday or Friday (לא בד"ו פסח), because as there is an invariable interval of 25 weeks and 2 days between the 1st of Nisan of the preceding year and the 1st of Tishri which follows (see table on p. 187), the interval between the 1st day of Passover (15th Nisan) of one year and the following New Year's Day must be 23 weeks and 2 days. Therefore Passover must always fall 2 days in the week earlier than the coming New Year. As New Year cannot fall on Sunday, Wednesday or Friday, therefore Passover cannot begin on Friday, Monday or Wednesday.

From the same rule it follows also that the first day of the other months, and hence also certain festivals or fasts, can only fall on certain days of the week.

[1] Yesod Olam iii., 15.

[2] Rosh Hashanah 20a. [3] Succah 43b.

3. Dechiyath Yach-Adu (דחית י"ח אד"ו) If as the result of the first rule (דחית י"ח), it would come about that 1st of Tishri would have to be postponed to Sunday, Wednesday or Friday, which are excluded by the 2nd rule (דחית אד"ו), then, obviously, the New Year's Day must be postponed for 2 days. Thus if the Molad occurs at or after noon on Saturday, Tuesday or Thursday, the New Year must be put off to Monday, Thursday or Saturday. Such a postponement is therefore called *Dechiyath Yach Adu* (דחית י"ח אד"ו).

4. Dechiyath Gatrad (דחית ג"ט"ר"ד). If in an ordinary (not leap) year, the Molad of Tishri is computed to fall on Tuesday between $11\frac{1}{3}$ minutes past 3 a.m. and 12 noon (i.e., between 3d. 9h. 11m. 20s. or 3d. 9h. 204ch., and 3d. 18h.), then New Year's Day is postponed till *Thursday,* because in such a case the Molad for the Tishri a year hence would fall 4d. 8h. 48m. 40s. later—that being the character of an ordinary year (see p. 191). But 3d. 9h. 11m. 20s. +4d. 8h. 48m. 40s. =7d. 18h. i.e. Saturday 12 noon, which by rule iii. (*Dechiyath Yach-Adu*) would necessitate postponing it till the following Monday, i.e., by 2 days. But as the Molad of the year in question fell on Tuesday before 12 noon therefore New Year began on Tuesday, and as the following year will fall on Monday, the year in question would be 356 days long which is more than the statutory limit of 355 days. (See Table p. 187.)

As 3d. 9h. 11m. 20s. =3d. 9h. 204 ch., this postponement is called *Dechiyath* $G_{a}t_{a}r_{a}d$ (דחית ג"ט"ר"ד). (see Chapter IV.).

5. Dechiyath B'tuthakpat (דחית ב"טו"תק'פ"ט). If the Molad of Tishri in any year following upon a leap year occurs on Monday after 9h. 32m. $43\frac{1}{3}$s. a.m. (i.e. 2d. 15h. 589ch.) the New Year must be put off to Tuesday. For, since the character of a leap year is 5d. 21h. 589ch., therefore the Molad of Tishri of that leap year must have fallen on 2d. 15h. 589ch. - 5d. 21h. 589ch. =9d. 15h. 589ch. - 5d. 21h. 589ch. =3d. 18h., i.e., Tuesday at noon, which means that the New Year's Day of that year must have been postponed till Thursday (by *Dechiyath Yach Adu*). Hence, if the current year were made to begin on Monday, the previous leap year would contain only 382 days, which, as seen from the Table on p. 187, is less than the lower statutory limit. (B'tuthakpat is composed of B = ב = 2d; Tu = ט"ו = 15h. Thak = תק = 500; P = פ = 80 and T = ט = 9).

Calculation of Duration of (i.e. number of days in) any given year.

(1) Divide the year (W.E.) by 19 to ascertain whether the year in question is an ordinary (פ) or leap year (מ).

(2) Compute the exact moments of the Molads for Tishri of the year in question and for Tishri of the following year, (pp. 190 and 191).

(3) Allow for the 5 postponements or *dechiyoth* and thus ascertain whether the year will be:

(a) Ordinary—containing 353, (ח), 354, (כ), or 355, (ש), days

or (b) Leap year—containing 383, ח) 384 (כ) or 385 (ש) days

(4) If the year contains 353, or 383 days, then the months of Marcheshvan and Kislev will contain* 29 days each and the year is defective (חסרה, or ח).

(5) If the year contains 354 or 384 days, then Marcheshvan contains 29 days and Kislev contains 30 days, and the year is regular (כסדרה, or כ).

(6) If the year contains 355, or 385 days, then Marcheshvan and Kislev contain 30 days each, and the year is full (שלמה, or ש)

Determination of the Calendar for the Year (*Kebiyah* קביעה).[1]

Having ascertained whether the year is an ordinary or leap year (see above), its calendar, or the days of the week on which the various feasts will occur are completely determined if we know the following facts about it, viz.:-

(1) The day of the week on which the 1st day of Tishri—or of the New Year—falls.

(2) The type of year—whether defective (ח), regular (כ) or full (ש). Thus, if we know that the year is an ordinary one, that the 1st of Tishri fall on a Monday (ב) and that the year is full (ש) then the formula for the year *Bash* (בש) completely determines its calendar. It is usual however to add another datum—in the form of a 3rd letter, viz.,

(3) The day on which the 1st day of Passover of that year falls. Thus *Bashah* (בשה) is the complete formula which states that New Year starts on Monday (ב), the year is full (ש) and the 1st day of Passover falls on Thursday (ה). These three-lettered formulae are called *Kebbiyoth* קביעות (fixtures) from *Kaba* (קבע) " to fix."

[1] The Talmud (Succah 48a), calls it קיבועא דירחא but this term probably refers to the Phase Method.

It is easy to find the 3rd letter in the *Kebiyah*, from the first two letters. For in an ordinary year, there are intervals of 2, 1 or 3 days in excess of a complete number of weeks between 1st Tishri and 1st day of Passover—according as the year is defective, regular or full (see Table on p. 187), so that if the day of the New Year, and the type of the year are given, the day on which Passover begins is at once known. Similarly, in the case of a leap year the 1st day of Passover must occur 4, 3 or 5 days later in the week than the first day of New Year according as the year is regular, defective or full.

Fourteen Types of Kebiyoth. We shall show that there can be only fourteen types of yearly calendars, seven for an ordinary and seven for a leap year, as follows:

Ordinary Year : *Bachag* (בחג), *Basha* (בשה) *Hakhaz* (הכז) *Gakha* (גכה)
(סימני הפשוטות) *Hasha* (השא) *Zacha* (זחה) *Zashag* (זשג)
Leap Year : *Bachah* (בחה) *Bashaz* (בשז) *Gakhaz* (גכז)
(סימני המעוברות) *Hacha* (החה) *Hashag* (השג) *Zachag* (זחג) *Zasha* (זשה)

In virtue of the varying length of a year, (between 353 and 355 days in the case of an ordinary year and between 383 and 385 in the case of a leap year), the 1st day of Tishri, i.e. New Year's Day will in any one year fall 3, 4 or 5 days later than the preceding ordinary year and 5, 6 or 7 days later than a preceding leap year. As, however, in virtue of Dechiyath Adu (see p. 193), the 1st of Tishri can only fall on a Monday (ב) Tuesday (ג), Thursday (ה) or Saturday (ז) we get the following kinds of year:

Types of Ordinary Years.

New Year Day	Length of Year	Day on which the New Year of the following year will fall		Possible or Impossible
		Calculation	Result	
Monday (ב)	Defective (ח)	3 days later	Thursday	Possible
	Regular (כ)	4 days later	Friday	Impossible (Adu p.193)
	Full (ש)	5 days later	Saturday	Possible
Tuesday (ג)	Defective (ח)	3 days later	Friday	Impossible (Adu)
	Regular (כ)	4 days later	Saturday	Possible
	Full (ש)	5 days later	Sunday	Impossible (Adu)
Thursday (ה)	Defective (ח)	3 days later	Sunday	Impossible (Adu)
	Regular (כ)	4 days later	Monday	Possible
	Full (ש)	5 days later	Tuesday	Possible
Saturday (ז)	Defective (ח)	3 days later	Tuesday	Possible
	Regular (כ)	4 days later	Wednesday	Impossible (Adu)
	Full (ש)	5 days later	Thursday	Possible

Hence we see that out of the 12 theoretically possible combinations only 7 are possible in practice, viz. :-

(1) בחג, (2) בשה (3) גכה (4) הכז, (5) השא (6) זחא (7) זשג

Similarly *in the case of a leap year* we get the following kinds of year:

Types of Leap Years.

New Year Day	Length of Year	Day on which New Year will fall in the following year — Calculation	Result	Possible or Impossible
Monday (ב)	Defective (ח)	5 days later	Saturday	Possible
	Regular (כ)	6 days later	Sunday	Impossible (Adu)
	Full (ש)	7 days later	Monday	Possible
Tuesday (ג)	Defective (ח)	5 days later	Sunday	Impossible (Adu)
	Regular (כ)	6 days later	Monday	Possible
	Full (ש)	7 days later	Tuesday	Impossible (See below)
Thursday (ה)	Defective (ח)	5 days later	Tuesday	Possible
	Regular (כ)	6 days later	Wednesday	Impossible (Adu)
	Full (ש)	7 days later	Thursday	Possible
Saturday (ז)	Defective (ח)	5 days later	Thursday	Possible
	Regular (כ)	6 days later	Friday	Impossible (Adu)
	Full (ש)	7 days later	Saturday	Possible

The reason that a *full* year beginning on a *Tuesday* is an impossibility is that even if we take the extreme time for the Tishri Molad of that year, viz., 3d. 17h. 1079ch., (since 3d. 18h. would entail the postponement of the new year to Thursday by the *Dechiyath Yach-Adu*) we shall have the Molad for the Tishri of the following year: 3d. 17h. 1079ch. + 5d. 21h. 589ch. (the latter being the *character* of a leap year) = 2d. 15h. 588 ch.—which being less than 2d. 15h. 589ch., the limit of the *Dechiyath B'tuthakpat* (p. 194), must make the New Year following the leap year fall on Monday and not on Tuesday. Hence we see that a leap year beginning on a Tuesday cannot be full, i.e., cannot have more than 384 days. We are therefore left with the following 7 *Kebiyoth* for a leap year:

(1) בחה (2) בשז (3) גכז (4) החא (5) השג (6) זחג (7) זשה.

Example. Determination of the Kebiyah for any given year (say 5689 W.E)

(i) $\frac{5689}{19} = 299 + \frac{8}{19}$

(1) $299(\underbrace{2d.\ 16h.\ 595ch.}_{\text{character of cycle}}) = 6d.\ 4h.\ 785ch.$

(2) $2(\underbrace{5d.\ 21h.\ 589ch.}_{\text{Charcter of leap year}}) = 4d.\ 19h.\ 98ch.$

(3) $5(\underbrace{4d.\ 8h.\ 876ch.}_{\text{character of ordinary year}}) = 7d.\ 20h.\ 60ch.$

(4) Molad Tohu (p. 190) = 2d. 5h. 204ch.

∴ Molad of Tishri 5689

= (1) + (2) + (3) + (4) = 7d. 1h. 67ch.

∴ First letter of Kebiyah is ז

(ii) Add " character " of 5689 (which is a leap year), 5d. 21h. 589 ch.

∴ Molad of Tishri 5690 = 5d. 22h. 656ch.

Hence by *Dechiyath Yuch-Adu* New Year of 5690 must be postponed to Saturday and the year 5689 begins on Saturday and ends on Friday thus consisting of 55 weeks (=385 days). It is therefore full (ש). Hence its *Kebiyah* is זש"ה.

The rebuke that Samuel received when he sent to Jerusalem a Calendar calculated for 60 years ahead (p. 12), as proof that the Jews in the diaspora were quite capable of making their own calendar without having to wait for messengers from Jerusalem, might make it appear that Samuel's Calendar was based on the method of mean conjunction containing nothing more than simple Arithmetical Manipulation which was within the competence of any mediocre mathematician[1] (חושבנא בעלמא ידע). In view, however, of the fact that on the basis of a Fixed Calendar, all that was necessary was to compute a Calendar for no more than 19 years (מחזור קטן), it would seem that Samuel's 60 year's Calendar was based on the True Conjunction Method — although with his t'kufah making the year 365d. 6h. long, his calculations must have been subject to slight error. The disfavour with which it was regarded was merely a reprimand for his attempt to sever the religious bond between the diaspora and Jerusalem. Indeed the early deaths of two of Samuel's daughters were attributed to the fact that it was due to him that a leap year was declared in Babylon without the authority of the Jerusalem Calendar Council.[2]

Calculation of T'kufoth[3]. The exact moment at which an equinox or a solstice occurred was of considerable importance during the Rabbinical period, since the question of intercalation

[1] Chulin 95 b. (2) Jer. Kethuboth ii. Quoted by A. Schwarz, op. cit. p. 32.
[3] Kidd. Hach. ix and x,

of an extra month (Adar II. or V'Adar) depended upon the date of occurrence of the vernal equinox (see p. 179).

(A) Samuel's T'kufah (תקופת שמואל). As a Samuel's year =365d. 6h.=52 weeks 1¼ days, the same t'kufah will occur 1¼d. later every year than in the preceding year, so that after 4 years it will occur at the same moment of the day, but 5 days later in the week, and after 28 (=4×7) years, it will recur at exactly the same moment of the day on the same day in the week. This *Twenty eight years' Cycle* is called the *Big* or *Solar Cycle* (מחזור חמה, מחזור גדול)[1], to distinguish it from the Small Nineteen Years' Cycle (מחזור קטן).

(a) To compute the day of the week and time of the day of any T'kufah.

(1) The Epoch is taken to be the *supposed* first T'kufath Nisan which is calculated to have occurred 4d. 0h. in the year 1 W.E. (Tuesday 6 p.m.)

(2) Divide the year in question—say 5689—by 28. The quotient giving the number of complete solar cycles elapsed since the epoch, may be neglected, since after 28 years the t'kufah recurs at the same moment of the day in the same day of the week. The remainder—5, in the case of 5689—represents the number which the year in question is in the current solar cycle. Hence the remainder less one (4 in our case) represents the number of years between the beginning of the current cycle and the t'kufath Nisan of the current year.

(3) Allow 1¼ days for each year of the current cycle (in our case 4×1¼=5 days) and add the product to the epoch. Therefore t'kufath Nisan of 5689=4d. 0h.+5d.=9d. 0h.=Sunday 6 p.m.

(4) As the interval between two successive t'kufoth=91d. 7½h. =13 wks. 7½h., therefore t'kufath Tammuz occurs 7½ hours later than t'kufath Nisan of the same year; t'kufath Tishri, 7½ hours later than t'kufath Tammuz and t'kufath Tebeth, 7½ hours later than the last.

(b) To compute the day of the month in which a T'kufah occurs, it is necessary to know (1) the day of the month in which the same t'kufah occurred in year 1, or the interval

[1] Berachoth 59b. (כל כ"ח שנין הדר מחזור ונפלה תקופת ניסן וכו')

between that t'kufah and the Molad of the immediately preceding or succeeding month; (2) The rate at which the t'kufah is delayed from year to year; (3) the rate at which the Molad of the month in question is delayed from year to year.

(1) The Molad of Nisan in the year 1, occurred on

$$\underbrace{2d.\ 5h.\ 204ch.}_{\textit{Molad Tohu}} + \underbrace{6(1d.\ 12h.\ 793ch.)}_{\textit{Charactor of month}} = \underbrace{11d.\ 9h.\ 642ch.}_{\textit{Wednesday 3h. } 35\frac{2}{3}\textit{m. a.m.}}$$

But t'kufath Nisan of year 1 occurred on the previous Wednesday at 0h. i.e., 4d. 0h. 0ch.

Therefore t'kufath Nisan *anticipated Molad Nisan* by 7d. 9h. 642ch.

Therefore t'kufath Nisan *followed Molad Adar* by

29d. 12h. 793ch. – 7d. 9h. 642ch., i.e., by 22d. 3h. 151ch.

(2) The number of days (in the month) that the t'kufah is delayed from year to year, i.e., the difference in dates between two successive t'kufoth of the same name is easily found, because:

(α) The **difference between a Solar** (Samuel's, or Julian) **year and ordinary year** of 12 lunations

= 365d. 6h. 0ch. – 12(29d. 12h. 793ch.)

= **10d. 21h. 204ch.**

(β) The **difference between a Solar year and a leap year** of 13 lunations = 365d. 6h. 0ch. – 13(29d. 12h. 793ch.)

= –**(18d. 15h. 589ch.)**

(γ) The **Excess of 19 Solar years over a 19 years' Cycle**

= 6939d. 18h. 0ch. – 6939d. 16h. 595ch.

= **0d. 1h. 485ch.**

Therefore in any given year which is N complete cycles, n_0 ordinary and n_1 leap years distant from year 1, the t'kufoth Nisan occurs 22d. 3h. 151ch. + N(0d. 1h. 485ch.) + n_0 (10d. 21h. 204ch.) – n_1 (18d. 15h. 589ch.) after the Molad Adar.

(3) The exact time of Molad Adar, of the year in question, is found in the manner explained on pp. 189 et seq. If then the value of the last expression turns out to be less than a lunation, i.e., less than 29d. 12h. 793ch., then the t'kufah falls in the same month, (i.e., in Adar). If it exceeds a lunation, then the t'kufah falls in the following month, (i.e., Nisan in the case of an ordinary year, and V'Adar in the case of a leap year).

The exact day in the month in which the t'kufah occurs is then ascertained by finding the interval between the Molad of the month in question (Adar, or V'Adar, or Nisan) and the first day of the month,—which is done by determining the K'biyah of the year.

Thus in the case of t'kufath Nisan of the year 5689, we have .
(22d. 3h. 151ch.) + 299 (0d. 1h. 485ch.) + 5 (10d. 21h. 204ch)—2 (18d. 15h. 589ch.)
= 57d. 7h. 288ch. ofter Molad Adar.
or 57d. 7h, 288ch.-29d. 12h. 793ch = **27d. 18h. 575ch.** after Molad V'Adar
(5689 being a leap year.)

But Molad of Tishri, 5689 = 7d. 1h. 67ch. (p. 191)
Therefore add 6(1d. 12h. 793ch.) = 2d. 4h. 438ch.
So that the Molad of V'Adar 5689 = 2d. 5h. 505ch.

Hence, (1) **Day of week, and time of day** on which t'kufath Nisan occurs in 5689, is 27d. 18h. 575ch. + 2d. 5h. 505ch. = 30d. 0h. = 2d. 0h. = **Sunday 6 p.m.**

(2) **Day in the Month** on which the t'kufah occurs is ascertained by finding how many days, after the Molad, the 1st day of V'Adar occurred in the given year. Now 5689 is not only a leap year (see p. 198) but calculation of its K'biyah (see p. 198), shows it to be a full year of which the 1st of Tishri is on Saturday (השז). Therefore the number of days between 1st of Tishri and 1st of V'Adar = 179 = 25 weeks and 4 days, (See Table p. 187) so that 1st of V'Adar falls on Wednesday, i.e., 4d. 0h. - 2d. 5h. 505ch. = 1d. 18h. 575ch. after Molad Adar.

Therefore **Day of Month** and time of day of the t'kufah in question
= 27d. 18h. 575ch. - 1d. 18h. 575ch. = 26 days after 1st of Adar = 27th V'Adar 0h. 0ch.

= **Sunday, 7th April, 1929, 6 p.m.**

(B) R. Adda's T'kufah[1]. R. Adda's year is exactly $\frac{1}{19}$ of a 19 years' cycle. Hence, *after every 19 years, the t'kufoth will recur* at exactly the same times of the day on exactly the same days of the week—in the same way as the t'kufoth Samuel recur every 28 years. Hence, **Cycle excess = 0d. 0h. 0ch.**

Also, (i) the **difference between a Solar (R. Adda's) and an ordinary leap year of 12 lunations** is

365d. 5h. 997ch. 48r. - 354d. 8h. 876ch.

= **10d. 21h. 121ch. 48r.**

(ii) the **difference between a Solar and a lunar leap year** = **—18d. 15h. 671ch. 28r.**

(iii) **T'kufah excess** = ¼ Adda's year—3 lunations

= **2d. 17h. 300ch. 31r.**

Hence, the time of any t'kufah can be easily calculated, if we know the time of that or any other t'kufah in any given year.

[1] Kidd. Hach, x.

The 1st *t'kufath Nisan*, according to R. Adda's system of calculation, occurred in *year I* W.E at 4d. 0h. 0ch. 0r. on the 29th of Adar, **9h. 642ch.** *before the Molad of Nisan* of the same year, *or*, which is the same thing, **29d. 3h. 151ch.** *after Molad Adar.*

Thus, as an example we shall find the date, day of week, and time of day, (for the meridian of Jerusalem) of t'kufath Nisan 5689. The year 5689 is the 8th year in the 300th cycle. Seven years elapsed in the current cycle, of which 5 were ordinary and 2 were leap years.

∴ T'kufath Nisan 5689.

=29d. 3h. 151ch. +5 (10d. 21h. 121ch. 48r.) - 2 (18d. 15h. 671ch. 28r.)

=46d. 5h. 496ch. 32r. after Molad Adar

=16d. 12h. 783ch. 32r. after Molad V'Adar.

But Molad V'Adar=2d. 5h. 505ch. 32r.

Hence **Day of week and time of day**=18d. 18h. 208ch. 32r.

=Wednesday 12h. 11m. 34s. p.m.

(2) Day of Month : As 1st of V'Adar fell on Wednesday, 2 days after the Molad (see p. 201), therefore 16 days after the Molad=14 days after the 1st=15th V'Adar.

Hence the t'kufath Nisan of year 5689 fell on Wednesday the 15th of V'Adar, 18h. 208ch. 32r. corresponding to Wednesday, 27th March, 1929, 12.11½ p m. (approximately)

NOTE : Comparison of the time of t'kufath Nisan according to R. Adda's system with that found according to Samuel's system of reckoning, thus shows a difference of approximately 11½ days. As a matter of fact, each of these systems is incorrect, since the exact length of the tropical year is less than that given by R. Adda. The exact length is 365d. 5h. 48m. 46·1s. (i.e. 365·2422001 days) (Samuel's year=365·25d. and R. Adda's=365·246822d.). The real date of the t'kufath Nisan or Vernal Equinox at the present time is about 21st March, i.e. 6 days earlier even than that calculated by R. Adda's system. Nevertheless the t'kufah Samuel is, on account of its greatest simplicity, the one still adopted by the modern Calender makers, and is the one recorded in the Jewish Almanacks, although the importance of the t'kufah from the point of view of Calendar making, is—at the present day—negligible.

Mahler suggests a t'kufah consisting of 91d. 7h. 489ch. 35r , making the tropical year equal to 365d. 5h. 877ch. 64r. =365 2422007 days=365d. 5h. 48m. 46 sec. which is almost identical with the actual present value of the tropical year of 365·2422001 days, the difference being about ½ second. (See p, 207).

Elucidation of a difficult Talmudic Passage.

We shall now show that the statement in the Palestinian Talmud cited on p. 75, that according to Rab the Saturnalia commenced 8 days before the Winter Solstice (T'kufath Tebeth) of the year 4009 W.E., whilst according to R. Yochanan, they began 2 days before the solstice,[1] can be explained on the supposition that Rab calculated by Samuel's system, and R. Yochanan used R. Adda's system.

[1] J. Abodah Zarah i. 2.

(1) Working by Samuel's system we have:-

Number of years between year 1 and year 4009 = 4008, containing 210 complete nineteen years' cycles, and 18 years—of which 12 were ordinary and 6 were leap years. Therefore, by formula on p. 200.

T'kufath Nisan (4009 W.E.) = 22d. 3h. 151ch. + 210(1h. 485ch.)
+ 12(10d. 21h. 204ch.) − 6(18d. 15h. 589ch.)
= 53d. 12h. 475ch., after Molad Adar,
Subtract one lunation (29d. 12h. 793ch.)
= 23d. 23h. 762ch., after Molad V'Adar,
Subtract one t'kufah excess i.e., ¼ Samuel's year—3 lunations
= 2d. 17h. 321ch
∴ T'kufath Tebeth (4009 W.E.)
= 21d. 6h. 441ch., after Molad Tebeth,

Now, Molad Tishri (4009 W.E.) can be shown, in the manner explained on p. 189 to have occurred on 3d. 19h. 960ch.; therefore by *Dechiyath Yach-Adu* (p. 194), 1st of Tishri of that year, occurred two days later, viz., on Thursday, and T'kufath Tebeth occurred on the 22nd day after Molad Tebeth.

We now have to find when the 1st day of Tebeth occurred in 4009, by determining the K'biyah for that year, as follows:-

Molad Tishri, 4009 = 3d. 19h. 960ch.
Add for leap year 5d. 21h. 519ch.
∴ Molad Tishri, 4010 = 2d. 17h. 469ch.

Therefore, by *Dechiyoth B'tuthakpat* (p. 194), 1st of Tishri, 4010, was postponed to Tuesday, making the duration of the leap year 4009 equal to 54 weeks + 5 days = 383 days. Therefore, the interval between 1st of Tishri and 1st of Tebeth, 4009, consisted of 88 days, or 12 weeks + 4 days (see Table on p. 187). But 1st of Tishri, 4009, fell on Thursday, therefore 1st of Tebeth, 4009, fell 4 days later, viz., on Monday.

Now, Molad Tishri, 4009, = 3d. 19h. 960ch.
Add 3(1d. 12h. 793ch.) = 4d. 14h. 219ch.
∴ Molad Tebeth. 4009, = 1d. 10h. 99ch.

Therefore, 1st of Tebeth, 4009, fell one day later than Molad Tebeth. Hence, *T'kufoth Tebeth* which occurred on the 22nd day after the Molad, *occurred on the 21st day of Tebeth, corresponding to 25th December*, 248 C.E. (see p. 205).

Therefore the Roman Saturnalia which begun on 17th December, started in the year 4009 W.E., or 248 C.E., *8 days before Samuel's Tkufath Tebeth.*

(2) Working by R. Adda's system, we have:-

T'kufath Nisan (4009 W.E.)=29d. 3h. 151ch.

+12(10d. 21h. 121ch. 48r.) – (6(18d. 15h. 671ch. 28r.)

– 29d. 12h. 793ch. = 18d. 6h. 29ch. 28r. after Molad V'Adar.

Subtracting T'kufah excess (2d. 17h. 300ch. 31r.), we get:

T'kufath Tebeth (4009)=15d. 12h. 808ch. 73r. after Molad Tebeth.

But Molad Tebeth, 4009, occurred on Sunday, and 1st of Tebeth fell on Monday (see p. 203). Therefore, *T'kufath Tebeth occurred on Monday, 15th Tebeth, corresponding to 19th December* 248 C.E.

Therefore the Saturnalia began 2 days before R. Adda's t'kufath Tebeth. The apparent contradiction between Rab and R. Yochanan is thus reconciled.

Correspondence between Jewish and Christian Dates.

Once the exact Jewish date of any event is known it is easy to find by means of special tables the corresponding date in the Christian Calendar. There is also a formula known as Gauss's Formula,[1] which may be utilised for the purpose, but the following arithmetical method my be used, **to find the date corresponding, say, to 21st of Tebeth, 4009.**

Molad Tohu (i.e., 2d. 5h. 204ch.) *corresponded to 6th Oct.* 3761 B.C. *11h.* $11\frac{1}{3}$ *p.m.* (p. 190).

Now, between Molad Tohu and Molad Tishri of year 4009 there elapsed 4008 years consisting of 211 complete 19 years' cycles, less one leap year of 13 lunations, amounting to

211(6939d. 16h. 595ch.)–383d. 21h. 589ch. = 1463890d. 14h. 756ch.

Add 11h. 11m. 20s. = 5h. 204ch.

∴ Time elapsed since 6th Oct., 3761 B.C., 6 p.m. =

1463890d. 19h. 960ch.

i.e., the Molad of Tishri, 4009 W.E. occurred on the 1463891st day after 6th October, 3761 B.C.

But 1st of Tishri, 4009, was, as we have seen, postponed by 2 days.

∴ Interval between 6th October, 3761 B.C., and Ist Tishri, 4009 = 1463892 days. Also as 4009 was a defective year (see p. 203), interval of 88 days, (see Table p. 187), making a total of 1463980 days.

[1] See Burnaby, *Elements of the Jewish and Muhammadan Calendars*, London 1901, p. 219.

Further, there was an interval of 20 days between 1st of Tebeth and 21st of Tebeth (i.e., day of t'kufah).

∴ Complete interval between October 6th, 3761 B.C., and t'kufath Tebeth, 4009 W.E. = 1464000 days, and the t'kufah occurred on the 1464001st day.

Now, 1 Julian year = 365·25 days, therefore 4 Julian years = 1461 days. Dividing 1464000 by 1461 we get a quotient of 1002 and a remainder of 78 days.

∴ Interval of time between 6th October, 3761 B.C., and date of t'kufath Tebeth, 4009 W.E = 1002 × 4 = 4008 Julian years, together with 78 days.

Subtracting 3761 years, we get 4008 − 3761 = 248, so that the year of the given t'kufah = 248 C.E., and the date of the t'kufah is 78 days after 6th October, i.e., the t'kufah occurred on the 79th day after 6th October = 25th December, 248 C.E.

The Differences between the Phase and Fixed Calendars are given in the following Table :-

Fixed Calendar	Phase Calendar
1. First of Month determined by Molad.	1. First of Month = Day of Sighted Moon.
2. Molad = mean conjunction, which may differ by about 15 hours from a true conjunction.	2. The visibility of the Moon was calculated.
3. Month is of fixed duration, viz., 29d. 12h. 44m. $3\frac{1}{3}$s. (Average Synodic Month).	3. Length of Month variable.
4. Months determined for any number of years ahead, and dates of festivals become automatically fixed, once the Molad of Tishri is known.	4. Beginning of a month is determined each time by observation and calculation.
5. Molad Tishri marks the beginning of a year, unless postponed on account of the various *Dechiyoth*. New Year's day cannot fall on Sunday, Wednesday, or Friday.	5. New Year was the day on which the new moon was sighted—whatever the day of the week.
6. The 19 years' cycle is used with the regular sequence of leap years (3, 6, 8, 11, 14, 17, 19). Length of Tropical year is taken as 365d. 5h. 55m. $25\frac{25}{57}$s.	6. The 19 years' cycle with its regular sequence of leap years was not used. A leap year was declared when the vernal equinox occurred after the 16th of Nisan.

The Various Eras in Jewish Chronology. The *World era* (W.E.) or *era of the Creation* (לבריאת עולם or ליצירה), whose epoch, as we have seen is the *Molad Tohu*, which occurred at 11h. 11⅓m. p.m. on what corresponds to Sunday, 6th October, 3761 B.C., is the one still used in all Jewish official ecclesiastical documents, such as Marriage Contracts (כתובות) Bills of Divorcement (גיטין) etc., and is frequently used by the very strict Jews in ordinary life. It was generally adopted in the 10th century C.E. The Talmud[1] refers to it when it states that 400 years after the *Destruction of the Second Temple* corresponds to the year 4231 of the World era.

אמר ר' חנינא אחר ארבע מאות לחורבן הבית... במתניתא תנא אחר ארבעת אלפים ומאתים ושלשים ואחת שנה לבריאת עולם, וכו'.

As 4231 W.E. corresponds to 4231 − 3761 = 470 C.E., this quotation dates the *destruction of the Temple* at 70 C.E., and this epoch formed another era (לחורבן)

In addition to these two eras, there was the *Seleucidan era* — named after Seleucus Nicator, a general of Alexander the Great, who conquered Gaza in 312 B.C., and its epoch is generally taken as October 1st of that year. It is also called the *Greek era*[2] (מנין יונים), or the *Era of Contracts*[3] (מנין שטרות) because legal documents were dated according to this era. It was superseded in the 10th century by the World Era.

During the Biblical period, various important historical events or unusual natural phenomena were taken as the epochs of new eras. Thus, Araphaxad, the son of Shem, was born two years *after the flood* (אחר המבול) (Gen. xi. 10). King Solomon built his Temple in the 480th year after *the Exodus* (I. Kings vi. 1). The Exodus era is also referred to in the Talmud,[4] where Rab Nachman, in reply to R. Acha bar Jacob, said that in the diaspora one reckons according to the Greek era and not according to the Exodus era

מתקיף לה רב אחא בר יעקב ממאי דלמלכות יונים מנינן דלמא ליציאת מצרים מנינן... אמר רב נחמן בגולה אין מונין אלא למלכי יונים.

A certain prophecy by Amos is stated to have been given two years before *the earthquake* לפני הרעש, (Amos i. 1), whilst a prophecy of Ezekiel is dated as the twenty-fifth year of the *Babylonian Captivity* לגלותנו (Ezek. xl. 1).

[1] Abodah Zarah, 9b. [2] Ibid, 10a. [3] Maimonides, Gerushin, 1-27.
[4] Abodah Zara, 10a.

Defects and Suggested Reform of the Jewish Calendar.

R. Adda's year of 365d. 5h. 55m. 25·44s. is about 6·6 minutes longer than the present value of the tropical year, viz., 365d. 5h. 48m. 46·09s., so that the accumulated difference during the 1500 years that the Fixed Calendar has been in use is $\frac{1500 \times 6}{24 \times 6}$ = about 7 days. Hence the beginning of Passover—which according to Biblical injunction should fall in the *first Spring month* (חדש האביב), viz., during the 30 days between 21st of March and 19th of April —has been shifted forward a complete week, and will, in time, regularly occur after the 19th of April. Indeed, even now, it sometimes happens that in a leap year Passover occurs later than in the first Spring month. Thus, in the year 5689 W.E. (=1929 C.E.) that festival began on the 25th of April, i.e., 6 days later than the Biblical limit. In order to remedy this it is necessary in the first instance to fix a more exact value for the length of the tropical year, such for instance as that suggested by Mahler (see p. 202); and secondly to introduce a different cycle with a different sequence of leap years. This can be done as follows:

Present value of synodic month = 29d. 12h. 44m. 3s. (p. 136).

∴ Present value of a lunar year = 12(29d. 12h. 44m. 3s.)
= 354d. 8h. 48m. 36s.

Length of Tropical year = 365d. 5h. 48m. 46s.

∴ Difference between these kinds of year = 10d. 21h. 10s.

This is equal to $\frac{\text{10d. 21h. 10s.}}{\text{29d. 12h. 44m. 3s.}} = \frac{939610}{2551443}$ of a synodic month. Converting this ratio into a continued fraction, we get:

$$\cfrac{1}{2+\cfrac{1}{1+\cfrac{1}{2+\cfrac{1}{1+\cfrac{1}{1+\cfrac{1}{17 \;\vdots\; +\frac{247}{7625}}}}}}}$$

If we stop at the dotted line, the value of the fraction becomes $\frac{123}{334}$ which means that very approximately the excess of a tropical year over a lunar year amounts to $\frac{123}{334}$ of a synodic month. Hence, if in the course of every cycle of 334 years we introduce 123 leap years, the excess would be practically wiped out. Indeed, such an arrangement would make the 334 civil years only about 39 minutes less than 334 tropical years, and it would take about 12,500 years (instead of about 200 years as it does now) for the difference to accumulate to one day.

As $\frac{334}{19} = 17$, together with a remainder of 11, the order of the leap years could continue as at present (see p. 189), for 17 complete "nineteen years' cycles," as well as for the first 11 years of the 18th cycle. After that a new "three hundred and thirty-four years' cycle" would have to begin again.

The Proposed Reform of the Secular Calendar and the Jewish Sabbath.

The League of Nations is now having under consideration the question of reforming the secular calendar. The *defects of the present Calendar* are :-

(1) The months are of unequal length—varying between 28 and 31 days.

(2) With the exception of February in ordinary years, a month consists of 4 weeks and 2 or 3 days. The months therefore begin on different days in the week, thus giving rise to considerable confusion in industrial affairs and sociological and other statistics. Thus, employees paid by the month sometimes have to work more and sometimes fewer days for the same pay. On the other hand, firms paying wages by the week have to make adjustments in their monthly accounts. Similarly, comparison of the amounts of business done in two different months must make allowances for the comparative number of working days in the two months—whilst even in the case of the same month in two different years, there may be more Sundays in one than in the other, since the same months do not start on the same day in the week in two consecutive years.

(3) The ordinary year consists of 365 days, which is one day more than a complete number of weeks (52).

The Proposed Reform is a very simple one. It is to make the year consist of **13** months, each one consisting of 28 days, and to make each month begin on Sunday and end on Saturday. As, however, $13 \times 28 = 364$ and a year consists of 365 or 366 days, it is further proposed to disregard these extra days—which would be called " blank days," or " year days," and would be inserted between Saturday the 28th of the last month of the year and the 1st day of the new year, which would be called Sunday.

Thus it is proposed to introduce this reform in 1933. The 1st of January, as well as of the other months of that year, will start on Sunday, and the 28th of the 13th month of that year will fall on Saturday (as every month will consist of just 4 weeks). The 365th day of the year will therefore fall on a day which we now call Sunday, but which, under the proposed scheme, will be called " blank day " or " year day." New Year's day of 1934 will therefore fall on what we now know as Monday—but will be called Sunday, and so on.

The Orthodox Jewish objection to the proposed reform is that the introduction of the blank days will result in a " travelling " Sabbath. For, as the Sundays in 1934 will be on days that are really Mondays, the Fridays will be on the days that are the Jewish Sabbaths, and Saturdays will be on the real Sundays. In the following year the 365th day will again be dropped, and observant Jews will have to observe Sabbath on what will universally be known as Thursdays. Similarly in 1936, Sabbath will have to be observed on Wednesday, and as that year is a leap year, the two last days will be blank and in 1937 Sabbath will fall on Monday.

This travelling Sabbath will cause great hardships to Orthodox Jews, because they will have to abstain from work on a different day in the week each year, as the ruling of the ecclesiastical authorities is that it is not so much the idea of resting one day in seven that matters from the religious point of view, as that that day should always be the same day in the week—the day which has always been known as Saturday. Several alternative schemes have, therefore, been proposed, of which the following seems to be the best. The year is to consist of 364 days, divided into 13 months of 4 weeks each. The 365th day in the ordinary year and the 366th day of every leap year should be accumulated, and when, after 5 or 6 years, they amount to 7 days, an extra week should be added to the year.

CHAPTER XVIII.

The Stars.

מונה מספר לכוכבים לכלם שמות יקרא (תהלים קמ"ז, ד')

" He telleth the number of stars ; He calleth them all by their names "

(Psalms cxlvii. 4)

The Stars are divided into (1) Fixed Stars, (2) Planets and (3) Comets. The last two belong to the Solar System, which is limited by the orbit of the outermost planet Neptune (whose mean distance from the sun is about 30 times as great as that of the earth), whilst the first belong to the Stellar Universe, consisting of untold millions of stars, each of which is comparable to our sun.

(1) **Fixed Stars** are those which, owing to their enormous distances from the earth—the nearest being so remote that light from them travelling at the rate of 186,000 miles a second takes about 4·3 years to reach us, and the farthest so far measured, being some 250,000 times as distant—so that we see them as they were 900,000 years ago!—appear to have no *proper motion* of their own. Halley in 1718, however, was the first to show that some of the stars—which the older astronomers believed to be fixed—do, after a very long interval of time, show a very slight, but perceptible, change of position in relation to one another. Thus Sirius, whose position in the heavens was carefully noted by Ptolemy, has, during the 1800 years that have since elapsed, moved $\frac{2}{3}^\circ$, i.e., at the rate of about $1\frac{1}{2}''$ a year—an angular velocity of about $\frac{1}{100,000}$ that of the sun. As Sirius is about 600,000 times as far from us as the sun, it follows that the actual velocity of Sirius is about the same as that of the sun—viz., roughly 1,000 miles a minute! Some of the more distant stars have been found to move at the rate of about $0{\cdot}04''$ per year, and, probably, those which have so far shown no movement at all are so remote that even the most refined instruments at present available are unable to detect it. Thus, in the case of a star 100 times as far as Sirius, even if it moved at the rate of 1,000 miles a minute, it would take 100 years to move $1''$ of arc—an angular rate of movement not very easy to detect with certainty at the present day, even by photographic methods.

The ancients, who were unable to measure the distances of stars, believed all of them to be equidistant from the earth and to be fixed on the surface of a sphere which surrounded those of the sun, moon and planets. They were therefore called *Fixed Stars.*

The Rabbis, in accordance with the Ptolemaic system, postulated the existence of seven spheres, but whilst Ptolemy's spheres contained the orbits of the moon, sun, and the five then known planets (Mercury, Venus, Mars, Jupiter and Saturn), outside which was an eighth sphere—the Stellar—the Talmudic sages, basing themselves on certain Biblical verses, assigned to these spheres certain metaphysical functions. Thus, Resh Lakish said,[1] there are seven firmaments as follows:

(1) Vilon (וילון), which is a curtain that hides the stars during the day.

(2) Rakia (רקיע), contains the orbits of the sun and moon.

(3) Sh'chakim (שחקים), contains the millstones for grinding manna for the righteous.

(4) Z'bul (זבול), accommodates Jerusalem and the Temple.

(5) Maon (מעון), the residence of angels.

(6) Machon (מכון), contains snow, hail, etc.

(7) Araboth (ערבות), contains, among other things, the souls of those yet unborn, as well as of the righteous departed ones.

According to R. Acha bar Jacob, there is still an eighth firmament—beyond which it is forbidden to penetrate for it is written in the book of Ben Sira. (190 B.C.)

במפלא ממך אל תדרש, ובמכסה ממך אל תחקר, במה שהרשית התבונן ואין לך עסק בנסתרות.

"Seek not the things that are too hard. for thee, and into the things that are hidden from thee enquire thou not. Instruct thyself in what is permitted to thee; thou hast no business with secret things."

Some idea of the **Talmudic Conception of the Immensity of Space** may be gathered from the following Allegory:[2] "When that wicked man (Nebuchadnezzar) said, 'I will ascend above the clouds; I will be like the Most High,'[3] he was rebuked by a heavenly voice (בת קול) saying: 'Thou wicked man, how many are the years of a man? Seventy! But the distance between the earth and the first firmament is a journey of 500 years, and a similar distance separates each sphere from the one next to it. Above these spheres are the holy living creatures (חיות הקדש)

[1] Chagigah 12b, 8a. [2] Chagigah 13a. [3] Isaiah xiv. 4.

each of whose 9 parts between the feet and the horns is of a measure corresponding to *all the preceding measures put together* and above these is the Throne of Glory whose feet are of a corresponding measure and whose seat is also of corresponding measure. God Almighty sits upon it! And yet thou didst say: I will ascend above the heights of the clouds, I will be like the Most High ' " !

If we sum all these measurements, we find the total amounting to a distance of 4,096,000 years' journey—which, at the rate of 80 miles in 24 hours (p. 92) amounts to 239,206,400,000—say 240,000 million miles— or about 2,600 times the distance of the sun. This is about one hundredth of what we now know to be the distance of the nearest fixed star. Nevertheless, in spite of the metaphysical way of expressing it, it is seen that the Rabbis had a fair idea of stellar distance.

Star Magnitudes. Stars are classified according to their degrees of brightness as being of the first, second, third, etc., magnitudes. The Talmud classifies them into 3 degrees of magnitude, viz., " Big stars that can be seen in the day time; small stars that can only be seen at night, and medium stars that are visible during twilight."[1] (See p. 93.)

The Number of Stars. The number of stars that it is possible to distinguish with the naked eye on a clear, moonless night, in the whole celestial sphere, is about 5,000. With the aid of powerful telescopes, however, the number which has been brought into view is about 1,500,000,000, or about one star for every living human being, which would almost seem to be a fulfilment of the promise God made to Abraham.[2] Dynamical considerations, however, make it practically certain that the number is at least 60 times as many and possibly more. The following Talmudic passage gives a hyperbolical estimate of the number of stars:[3] " Each of the twelve Mazoloth (signs of the zodiac) has 30 armies (חיל). Each army has 30 legions (לגיון) each of which has 30 divisions (רהטון). and each of these has 30 cohorts (קרטון). Each cohort contains 30 camps (גסטרא), each of which comprises 365,000 myriad stars." A literal interpretation would make this estimate amount to $12 \times (30)^5 \times 365 \times 10^7 = 106,434 \times 10^{13}$ say 10^{18}— i.e., nearly a thousand million times as many as have so far been counted by means of the most powerful telescopes. The following lines of Mr. Allingham, quoted

[1] Sabbath 35b. [2] Gen. xv. 5. [3] Berachoth 32b.

by Sir Robert Ball,[1] express the countlessness of the stars even more forcibly and perhaps more extravagantly than the Talmudic passage just quoted:-

"But number every grain of sand,
Wherever salt wave touches land;
Number in single drops the sea;
Number the leaves on every tree,
Number earth's living creatures, all
That run, that fly, that swim, that crawl;
Of sands, drops, leaves and lives, the count
Add up into one amount.
And then for every separate one
Of all these, let a flaming sun
Whirl in the boundless skies, with each
Its many planets to outreach
All sight, all thought: for all we see
Encircled with infinity,
Is but an island."

Constellations. The various stars were grouped into constellations, to which the ancients gave fanciful names. To the names of the signs of the zodiac we have already referred (p. 79). The northern constellation, the Great Bear (עש) or the Waggon (עגלה) is mentioned[2] as "being never seen in the southern skies" and certain myths are told about the Pleiades (כימה) and Orion (כסיל)[3] The Talmud puts the number of stars in Pleiades at 98. The actual number that the modern telescope reveals in that cluster is over 600, but in view of the fact that the average unaided eyes can see only 7 stars, that people with unusually acute vision can see no more than about a dozen, and that even Galileo, with his telescope, could not discern more than 40, the estimate given by the Rabbis is a very remarkable one.

The Milky Way, which we now know to be made up of countless numbers of stars of less than the 8th degree of magnitude, was first described by Aristotle, although he was not aware of its nature. It is possibly referred to in Daniel, vii, 10, as the "fiery stream" נהר די נור, and expatiated on in the Talmud,[4] where its origin is ascribed to exudations of the mythological "living creatures." "The sting of Scorpio," says the Talmud, "may lie in the Milky Way."[5] עוקצא דעקרבא דמנח בנהר דינור.

(2) **The Planets** (כוכבי לכת), are star-like bodies which have orbital motions of their own. There are many planets known at the present day and their number is continually being added to by new discoveries. At the time of the Talmud, however, and indeed, till the end of the 18th century, when Herschel discovered Uranus,

[1] "The Story of the Heavens," London 1891.
[2] Pesachim 94b. [3] Berachoth 58b. [4] Chagigah 13b. [5] Berachoth 58b.

astronomers were acquainted with only five planets—in addition to the sun and moon, which were also classed as planets—each one moving in its own orbit round a fixed earth. Their order from the earth outwards was: the Moon (לבנה) Mercury (כוכב), Venus (נוגה), the Sun (חמה) Mars (מאדים) Jupiter (צדק) and Saturn (שבתי). The Talmud[1] enumerates these and describes the supposed influences they have on the future career of man. Rashi ad loc. gives a mnemonic for them (borrowed from Pirke' R. Eliezer), formed from the initials of their names—from without inwards SheTZaM CheNCaL (שצם חנכל).

That the planets were not all at the same distance from the earth was inferred from the fact that certain planets were sometimes eclipsed by the moon (a phenomenon called **occultation)** indicating that the moon was the nearest of them all. The relative distances of the other planets were estimated from their relative periodic times: the less the periodic time, the smaller the distance from the earth. The Talmud, however, does not give the distances. Occultations are referred to by Isaiah (xiii, 10; see p. 127). Job (ix, 7) also alludes to them: " He sealeth up the stars " ובעד כוכבים יחתום.

The Periodic Times of the planets are not given in the Talmud, but are mentioned in the Midrash.[2] The following Table compares the periodic times of the planets as given in the Midrash with those obtained from observation and calculation (by means of Kepler's 3rd law, see p. 154):-

Planet	Periodic Time	
	Midrash	Calculation
Mercury (כוכב)	480 years	87·97 days
Venus (נוגה)	10 months	224·69 days
Mars (מאדים)	$1\frac{1}{2}$ years	687·00 days
Jupiter (צדק)	12 years	11·87 years
Saturn (שבתי)	30 years	29·44 years

The actual text of the Midrash assigns a periodic time of 480 years also to Venus and Mars, but a gloss corrects the figures to those given in the Table. The periodic time of Mercury, however, is for some reason left unaltered. The Midrash also says that until Adam sinned, the planets moved in smaller orbits. That the Talmudic sages were able to calculate the orbits of the planets is evident from this statement: It is reprehensible for one who is able to do these calculations not to practice this art[3] כל היודע לחשב בתקופות ומזלות ואינו חושב, וכו'.

[1] Sabbath 156a. [2] Gen. R. x. 43. [3] Sabbath 75a.

Comets. It is generally believed that Cochba de Shebit[1] refers to a comet, although Brodetsky[2] is of opinion that it designates a shooting star. The Physical nature of comets and their orbits was unknown until Newton showed them to move in ellipses or in parabolas—the periodic ones in the former and those which are seen once only, and never return, in the latter. It is, therefore, not surprising that the great Talmudic astronomer Samuel acknowledged his ignorance of the nature of these celestial bodies.[3]

נהירין לי שבילי דשמיא כשבילי דנהרדעא לבר מכוכבא דשביט דלא ידענא מאי ניהו.

It is, however, remarkable that although, until the middle of the 17th century, even the most civilised nations did not know the period of any comet, the periodic time of at least one comet was known to R. Joshua in the second century, for he said : כוכב אחד לע' שנה עולה ומתעה את הספנים. " There is a star which appears once in 70 years and misleads the captains of boats." Regarding the alleged identity of this with Halley's Comet, see p. 11.

Comets have from the earliest times—until their nature and orbits were discovered—been regarded as the harbingers of wars, famines and other dire events. Josephus[4] mentions that " Amongst other warnings (of the destruction of Jerusalem in 70 C.E.), a comet, of the kind called Xiphias, because their tails appear to represent the blade of a sword, was seen above the city for the space of a whole year." It will be noted that the words of Josephus are almost identical with those used by King David about 1,000 years earlier under similar circumstances.[5] " And God sent an angel unto Jerusalem to destroy it . . . . And David lifted up his eyes, and saw the angel of the Lord stand between the earth and the heaven, having a drawn sword in his hand stretched over Jerusalem . . . . And David offered burnt offerings and peace offerings . . . . And the Lord commanded his angel and he put up his sword again into the sheath thereof."

Shooting Stars and Meteorites. The Talmud[6] states that at the death of R. Chiya, stones of fire fell from heaven.

כי נח נפשיה... דרבי חייא נחיתו כיפי דנורא מרקיע.

[1] Berachoth 58b. [2] Jewish Review, July 1909, p. 168. [3] Berachoth 58b.
[4] The Wars of the Jews, vi. 5. [5] I. Chron. xxi. 15to 28. [6] Moed Katan 25b.

Astrology. The subject of astrology, which deals with the supposed influence of heavenly bodies upon the careers and destinies of man, is one of considerable interest, but cannot be dealt with extensively in this book.[1] But, to render the account of Rabbinical astronomy complete, it is necessary to refer to the matter briefly. The ancient Babylonians—some 5,000 years ago—on the strength of written records of certain past events, good or bad, which happened to synchronise with the appearance of certain unusual astronomical phenomena, such as eclipses, occultations, the appearance of comets, etc., endeavoured to find some correlation between them, and gradually claimed to be able to predict future events from the appearance of the heavens. This belief was greatly elaborated by the Greeks, who originated the practice of tracing the Horoscope of a person from the position of the heavenly bodies at the time of his birth. This superstition flourished during the Middle Ages. (R. Abraham bar Chiya of the 11th century was a great believer in it and called it הנסיון). It has to some extent survived even to the present day.

Owing to its Babylonian origin, those who practised the art of Astrology were called by the Rabbis *Chaldeans* כלדאי, and some Rabbis were firm believers in the truth of this art. Thus Raba of the 4th century said that " the length of a man's life, the characters of his children, and even his means of livelihood depend upon the planets "[2]

חיי בני ומזוני לא בזכותא תליא מלתא אלא במזלא תליא מלתא.

Even Samuel, the expert astronomer and physician, considered it dangerous to do venesection on Monday, Tuesday and Thursday.[3]

The following " Fish Story " is interesting. The Astrologers once told a Gentile that his fortune would fall into the hands of a pious, Sabbath-observing, Jew. He therefore, exchanged his fortune for a large diamond, which he always carried with him. One day he accidently dropped it into the river, and it was subsequently found inside a fish, by a Jew who bought it for his Sabbath meal.[3]

According to R. Chanina, the planets exercise an influence, for good or evil, over the birth of individuals.[5] " One born under

[1] For greater detail see the article " Astrology " in the " Jewish Encyclopædia "
[2] Moed Katan 28a. [3] Sabbath 129b. [4] Sabbath 119a. [5] Sabbath 156a.

Venus will be rich and sensual; under Mercury—beautiful and wise; under Saturn—will have his resolutions come to nothing; under Jupiter—righteous and jovial; under Mars—he will shed blood, i.e., he will be either a murderer, a surgeon or a butcher." Other rabbis condemned such superstitions: "Away," said they, "with your astrology, for Jews are not subject to the influence of the planets " צא מאצטגנינות שלך שאין מזל לישראל.

It is interesting to note that R. Abraham bar Chiya, in a book of his called M'gillath HaM'gallé, מגלת המגלה, predicted the coming of the Messiah in the year 5018 W.E. (i.e., 1257-1258), "when Jupiter and Saturn will be in conjunction in the sign of Pisces." As Filipowski[1] remarks, had the author lived in that year he would have explained the non-fulfilment of the prophecy on the ground that the Jews had not yet deserved such an event, i.e., "the fault is not in the stars, but in ourselves."

EPILOGUE

We have now completed our survey of Rabbinical Astronomy. The Talmudic sages, although they fostered a study of astronomy for calendric purposes, looked with disfavour upon any attempt to "ascend above the heights of the clouds" and to ponder too deeply upon the mysteries of the heavens (p. 213). To R. Zoma, who lost his reason as the result of excessive speculation in the domain of astronomical mysticism, the Rabbis[2] applied the verse in Proverbs (xxv., 16): "Hast thou found honey, eat as much as is sufficient for thee, lest thou be filled therewith and vomit it" דבש מצאת אכול דייך פן תשבענו והקאותו. It is true that King Solomon said (Proverbs xxv., 2), that it is the glory of man to search out the secrets of Nature כבוד אלהים הסתר דבר וכבוד מלכים חקר דבר, but Ben Sira, a few hundred years later, discouraged such research, for he said, "thou hast no business with secret things" (p. 212). It is, therefore, no wonder that very few, if any, epoch-making discoveries in the field of mathematics and astronomy can be traced to the Jews of the Talmudic period.

[1] Abraham bar Chiya *Sefer Haibbur* edited by A. Filipowski, London, 1851.
[2] Chagigah 14b.

A superficial comparison between the immensity of the universe and the apparent insignificance of man calls to mind the sublime words of the Psalmist, who—in a moment of ecstasy—exclaimed:[1]

כי אראה שמיך מעשה אצבעותיך ירח וכוכבים אשר כוננתה מה אנוש כי
תזכרנו ובן אדם כי תפקדנו.

which may be freely translated, in the words of F. T. Palgrave,

> "Who is man and what his place
> Anxious asks the heart perplext,
> In this recklessness of space
> Worlds with worlds intermixt?
> What has he this atom creature
> In the infinitude of Nature?

But much as the contemplation of celestial things may humble us, it cannot help elevating us and giving us a sense of pride at the same time. For here is man—a minute but mighty atom—confined to a small spot on this tiny planet, and whose existence is limited to a few courses of the sun, but who has, nevertheless, an imagination which, on the one hand has devised means of unfolding the most complicated inner structure of the infinitesimally small chemical atom, and on the other, has with magnificent daring plunged into the boundless abyss of space and has thrown out a net which has entrapped within its meshes countless millions of stars, some of which are so remote that light leaving them at the rate of over 10 million miles a minute takes millions of years to reach us![2] With God-like skill man "hath measured the waters in the hollow of his hand, and meted out heaven with the span, and comprehended the dust of the earth in a measure, and weighed the mountains in scales and the hills in a balance."[3] When we realise these wonderful achievements of the human brain, have we not a right to exclaim with Hamlet: "What a piece of work is man, how noble in reason, how infinite in faculties . . . . . in apprehension how like a God!" and with the Psalmist (viii., 6)
ותחסרהו מעט מאלהים וכבוד והדר תעטרהו (תהלים ח' ו'.)

[1] Psalms viii, 4, 5.

[2] Some of the Spiral Nebulæ are estimated to be over 100,000,000 light-years distant.

[3] Isaiah xl. 12. מי מדד בשעלו מים ושמים בזרת תכן, וכו'

Summary and Conclusions.

In this book an attempt has been made to explain the various mathematical and astronomical matters mentioned in the Talmud, and to investigate the extent of the mathematical and astronomical knowledge of the ancient Jewish sages as is revealed from a study of the Babylonian and Palestinian editions of that astounding work and its principal commentators. I believe that we are justified to draw the following conclusions:-

(1) As regards astronomy, there is enough evidence to show that, although there was no Hipparchus or Ptolemy amongst them, the Rabbis certainly produced several astronomers of considerable merit, e.g., R. Gamliel (who possessed a kind of telescope), Samuel, R. Joshua (who appears to have known the periodic time of a comet), and a few others (see Chapter III.).

(2) As regards mathematics, one can, I believe, say quite conclusively that:-

(i) The ancient Jewish Rabbis were in respect of their mathematical knowledge more like the Babylonians, Egyptians and Romans, but certainly did not equal the Greeks in their geometrical speculations. Not only did they not produce such mathematical geniuses as Thales, Pythagoras, Euclid, Archimedes, etc., but there is no evidence whatever that they discovered anything original in mathematical theory. Indeed, their mathematical knowledge was of a very meagre and unsatisfactory kind. For, as we have seen, even such geometrical facts as the relations between the area of a circle and that of its inscribed or circumscribed squares, were obviously taken for granted (without any attempt at verification) *on the strength of the opinion of Roman philosophers* (see p. 25)—who were themselves very poor mathematicians.

(ii) The whole of the mathematics (which principally consisted of geometry) found in the Talmud is entirely of a practical character and was merely used by the Talmudic Rabbis as a help in the exposition and development of the civil and religious laws. In other words, whilst the Greeks studied mathematics for its own sake and felt a strong craving to speculate and discover new mathematical facts, the ancient

> Jews do not appear to have taken any interest in the development of mathematical theory, and were satisfied with applying, with a variable amount of skill, what simple mathematical tools they possessed to the various practical problems with which they had to deal.

Voltaire wrote: " Les juifs ne furent jamais ni physiciens, ni géomètres, ni astronomes." However untrue such an accusation would be at the present day in the face of such intellectual giants as Ehrlich, Widal, Wasserman, Schick, etc., in medicine, and Jacobi, Sylvester, Levi-Civita, the Herschels, Michelson, Hertz and Einstein in the field of mathematics, physics and astronomy, one cannot help agreeing with Voltaire, that, with the exception of Hygiene and Public Health founded by Moses, the foundations of these sciences were laid without Jewish co-operation.

APPENDIX ON TALMUDIC METROLOGY.

A. Weights

		Metric	English
Gerah (גרה), or Ma'ah (מעה), the Tyrian Coin, whose weight, known from existing specimens		= 6·0 grm.	= 9.26 grains
Zuz (זוז), *Zin*, (זין) or *Denarius* (דינר)	= 6 Gerahs	8·6 ,,	55·56 ,,
Common Shekel (שקל פשוט) or simply, Shekel	= 2 Zuzin	7·2 ,,	111·52 ,,
Sanctuary Shekel (שקל הקדש) or *Sela* (סלע)	= 2 Shekels	14·3 ,,	222·24 ,,
Tartimar (תרטימר)	= 25 Shekels	180·0 ,,	$6\frac{1}{3}$ ozs.
Italian Manah (מנה איטלקי) or Litra (ליטרא)	= 2 Tartimars	360·0 ,,	$12\frac{2}{3}$,,
Ordinary Manah (מנה)	= 1·6 Italian Manah	576·0 ,,	1 lb. 4·28 ozs.
Kikkar, or *Talent* (ככר)	= $37\frac{1}{2}$ Manahs = 3,000 Shekels	21·6 Kgm.	47·52 lbs.

NOTES: (1) The Biblical shekel contained 20 Gerahs (Exod. xxx), and Josephus (Ant. xiv). gives its weight as $\frac{1}{20}$ Roman pound = 16·37 grms. = 252 grains. The Kikkar, as seen from Exod. xxxviii. contained 3,000 Shekels — since the tax of ½ shekel per head (בקע לגלגלת) levied upon 603,250 men, i.e. 301,775 shekels, amounted to 100 kikkars and 1775 shekels; ∴ 100 kikkars = 300,000 shekels whence 1 kikkar = 3,000 shekels.

(2) The Talmudic weights differed in different parts of the Country.

(3) These weights in Silver or Gold were used as coinage. The values of such coins therefore varied with the value of these metals.

B. Measures

(a) **Linear:** The *finger-breadth* (אצבע) was of 3 kinds, viz.

(1) Thumb-breadth (גודל) 2·33 cm.
(2) Middle-finger breadth (תילתא) 1·87 ,,
(3) Little-finger-breadth (קטנה) 1·56 ,,

		Metric	English
The unspecified *finger-breadth* refers to a thumbreadth		= 2·33 cm.	= 0·93 inch
Handbreadth (טפח) = 4 thumbs,	= 5 middle fingers = 6 little fingers	9·34 ,,	$3\frac{3}{4}$,,
Hasit (הסיט)	= 2 handbreadths	18·68 ,,	$7\frac{1}{2}$,,
Span (זרת)	= 3 handbreadths	28·01 ,,	$11\frac{1}{4}$,,
Cubit (אמה), or *Pace* (פסיעה)	= 6 handbreadths	56·02 ,,	1 ft. $10\frac{1}{2}$ in.
Stadium (ריס)	= 1,600 ,, ,,	149·38 metres	162·82 yds.
Rabbinical Mile (מיל) or, *Sabbath Distance* (תחום שבת)	= $7\frac{1}{2}$ Stadia = 2,000 cubits	1120 4 metres	0.7 miles
Parsa (פרסא)	= 4 Rabbinical Miles = 8,000 cubits,	4481·6 ,,	2·8 ,,
[Parsa is also a Persian Mile	which = 12,000 cubits	6722·4 ,,	4·2 ,,][1]
Day's March (דרך יום)	= 10 Parsas	4481·6 ,,	28 ,,

[1] See p. 102.

Measures continued

(b) Superficial.

The only measurement of area given in the Talmud is *Beth Sothaim* (בית סאתים), i.e., the area of a field in which two se'ahs can be sown. This is equivalent to the area of the court in front of the tabernacle (חצר המשכן) $=100$ cubits long by 50 cubits wide $=5{,}000$ square cubits $=5{,}000\times(56{\cdot}02)^2=1569$ square metres $=$ 1875 square yards, i.e., approximately $\frac{3}{8}$ acre.

(c) Cubical.

(α) *Solids.*

Beitzah (ביצה) = volume of an average egg, as ascertained by the Archimedean principle (i.e., the volume of water it displaces).[1] Modern determinations make it equal to about 60 cc., but the Rabbis apparently made it equal to about $1\frac{1}{2}$ times this quantity, because the beitzah $=\frac{1}{144}$ se'ah, which latter $=\frac{1}{40}$ Mikvah. But a Mikvah $=3$ cubic cubits (p. 19) $=3\times(56{\cdot}02)^3=527413$ cc., so that a

beitzah $=\dfrac{527413}{144\times 40}$ = 91·56 cc.

Log (לוג) or *Kesta* (קסתא) = 6 beitzahs	=	549 cc.
Kapiza (קפיזא) = 2 Logs	=	1098 cc.
Kab (קב) = 2 Kapizas	=	2196 cc.
Se'ah (סאה), or Geriva (גריוא) = 6 Kabs	=	13176 cc.
Epha (איפה) or *Bath* (בת) = 3 Se'ahs	=	39528 cc.
Letech (לתך) = 5 Ephas	=	197640 cc.
Cor (כור) = 2 Letechs	=	395280 cc.
	=	395·28 litres

(β) *Liquids.*

Korteb (קורטב) $=\frac{1}{64}$ log $=\frac{549}{64}$ cc.	=	8·58 cc.
Barzinah (ברזינא) = 2 Kortebs	=	17·16 cc.
Antel (אנטל) = 5 Ephas	=	137·35 cc.
Log (לוג) = 4 Antels	=	549 cc.
Cuza (כוזא) = 5 Logs	=	3296 cc.
Metarta (מטרתא) = 12 Cuzas	=	39552 cc.
Mikvah (מקוה) = 40 Se'ahs	=	527040 cc.
	=	527·04 litres
	=	1165 gallons

[1] Kelim xvii. 6 .נותן לתוך המים וחולק את המים

INDEX OF RABBINICAL REFERENCES.

TALMUD.

MAIMONIDES.

SUBJECT INDEX

A

B

C

D

E

F

G

H

I

J

K

L

M

N

O

P

Q

R

S

T

V

W

Y

Z

ERRATA AND ADDENDA.

Page 4 lines 9 and 11, read "Systems" and "References."

Page 8 The Karaites still determine the New Moon by Observation.

Page 9 line 4 from bottom, read "Harchinas."

Page 13 lines 4 and 5, read "(see p. 70)."

Page 17 footnote 2, read "*γραμματεια*" and "*γεωμετρια*"

Page 23 line 4 from bottom, read "(12th century)."

Page 24 line 3 from bottom, read "if *we* accept."

Page 25 Footnote 1. Omit the reference to Herod's Capture of Cæsarea.

Page 35 line 10 from bottom, read "corners" instead of "periphery."

Page 38 line 6, read "suburbs."

Page 44 line 8, read "and one angle" instead of "and their contained angle."

Page 47 last line, read "(see Fig. 17)."

Page 52 lines 8 and 9 from bottom, read "diagonal" instead of "diameter."

Page 54 line 12, read "1 $\frac{3\frac{1}{2}}{16}$" instead of "$\frac{3\frac{1}{2}}{16}$"

Page 58 line 24, add after "says" the reference "(Erubin 78a)."

Page 63 line 7 from bottom, read "actually."

— For the sun and moon to be in conjunction, it is not necessary for their centres to be in the same straight line as the centre of the earth; it is only necessary for the sun and moon to have the same longitude (see p. 122).

Page 72 end of 3rd paragraph, omit the word "and."

Page 74 last paragraph, read "91d. $7\frac{1}{2}$h. × 4 = $365\frac{1}{4}$days."

Page 75 line 7, read "Hai Gaon" instead of "Hai Gain."

Page 80 line 4, read "Psalms cxlviii., 6."

Page 90 line 7, read "p. 82" instead of "p. 7."

Page 91 Note. It is only between March and September that the day is now gradually diminishing. Between September and March, when the sun is South of the equator, the day is lengthening. But the whole point is of very little importance.

Page 93 Other causes of variation in the apparent length of twilight are meteorological conditions, such as mist, humidity, as well as proximity or otherwise of mountains.

Page 102 line 3 from bottom. Although a Persian mile is 4·14 English miles, the Rabbinical Parsa was only $\frac{2}{3}$ of the Persian mile, or 2·8 English miles (see Appendix). This would make the Rabbinical measurements of the earth 33% too low.

Page 103 line 3, read "$\frac{50}{3}$" instead of "$\frac{1}{50}$"

Page 107 The remarks about the apparent motion of the sun being faster in the summer than in the spring, and faster in the autumn than in the winter, only hold good for the particular value of KEV shown in Fig. 40, which was the case in the time of Ptolemy. At the present day, owing to the gradual increase of the angle KEV, the sun's motion is fastest in winter and slowest in the summer (see Note on p. 110). Actually it is fastest on January 3rd (1°1′8″ per day) and slowest on July 6th (0°57′11″).

Page 109 Paragraph 1 gives the principle only of fixing the apse line, since the actual daily variations in the apparent diameter of the sun, being less than half a second, they cannot, even now, be accurately measured. Actually, the apse line is fixed by consideration of the sun's motion in longitude over long intervals of time.

Page 110 line 2 of Note, read " speed of the sun in the ecliptic."

Page 112 lines 1 and 17, read " revolution " instead of " rotation."

Page 124 line 6, read " $708\frac{2}{3}$ " instead of " $7384\frac{2}{3}$."

— line 8 from bottom, read " non-luminous " instead of " opaque."

Page 136 Table, under Duration of Synodic Month, read " 29d. 12h. etc.," instead of " 29d. 13h. etc."

Page 151 Table, read 0°59′ instead of 0°59″.

Page 169 line 9, read " $\tan \Upsilon D_1 = \tan \Upsilon C \cos 23\frac{1}{2}°$
$= \tan 30° \cos 23\frac{1}{2}° = \cdot 5774 \times \cdot 9171 = \cdot 5295$

Page 171 lines 4 and 7 read " see Tables p. 151."

Page 174 1st paragraph line 6, read " from his mind."

Page 176 line 11, read " Area of Crescent."

Page 181 last paragraph. It is possible for the moon's latitude to be north and yet for the moon to be left of the sun, but if the moon's latitude is south, it must be to the left of the sun (in the northern geographical latitudes), just before setting.

Made in the USA
Monee, IL
03 August 2026

58903149R00152